Thermodynamik

Energie • Umwelt • Technik

Band 17

λογος

Thermodynamik
Energie • Umwelt • Technik

Herausgegeben von Professor Dr.-Ing. Dieter Brüggemann
Ordinarius am Lehrstuhl für Technische Thermodynamik und
Transportprozesse (LTTT) der Universität Bayreuth

Numerical Modeling and Simulation
of PEM Fuel Cells under Different
Humidifying Conditions

Von der Fakultät für Angewandte Naturwissenschaften

der Universität Bayreuth

zur Erlangung der Würde eines

Doktor-Ingenieurs (Dr.-Ing.)

genehmigte Dissertation

vorgelegt von

Luis Matamoros, Master in Chemical Engineering

aus

Caracas (Venezuela)

Erstgutachter: Prof. Dr.-Ing. D. Brüggemann

Zweitgutachterin: Prof. Dr. M. Willert-Porada

Tag der mündlichen Prüfung: 10. Juni 2008

Lehrstuhl für Technische Thermodynamik und Transportprozesse (LTTT)

Universität Bayreuth

2009

Thermodynamik: Energie, Umwelt, Technik
Herausgegeben von Prof. Dr.-Ing. D. Brüggemann

Luis Matamoros:
Numerical Modeling and Simulation of PEM Fuel Cells under Different
Humidifying Conditions;
Bd. 17 der Reihe: D. Brüggemann (Hrsg.), „Thermodynamik: Energie,
Umwelt, Technik";
Logos-Verlag, Berlin (2009)
zugleich: Diss. Univ. Bayreuth, 2008

Bibliografische Information der Deutschen Nationalbibliothek

Die Deutsche Nationalbibliothek verzeichnet diese Publikation in der
Deutschen Nationalbibliografie; detaillierte bibliografische Daten sind
im Internet über http://dnb.d-nb.de abrufbar.

ISSN 1611-8421
ISBN 978-3-8325-2174-5

Logos Verlag Berlin GmbH
Comeniushof, Gubener Str. 47,
10243 Berlin
Tel.: +49 030 42 85 10 90
Fax: +49 030 42 85 10 92
INTERNET: http://www.logos-verlag.de

Vorwort des Herausgebers

Bei der Entwicklung umweltverträglicher Konzepte zur Energiewandlung findet die Brennstoffzelle für manche Anwendungsbereiche seit einiger Zeit und unvermindert großes Interesse. Bei nüchterner Betrachtung stellt man neben unbestrittenen Vorzügen jedoch auch technische und ökonomische Herausforderungen fest, deren Meisterung für eine breitere Markteinführung entscheidend ist.

Eines der technischen Probleme, welches bei PEM-Brennstoffzellen je nach Betriebsweise von Bedeutung ist, stellt der miteinander verknüpfte Wärme- und Wasserhaushalt dar. Sowohl eine zu hohe wie auch eine zu niedrige Feuchte beeinträchtigt die Funktionstüchtigkeit eines Brennstoffzellensystems gravierend.

Der Autor des vorliegenden Bandes hat sich damit beschäftigt, zunächst geeignete Modellvorstellungen zusammenzustellen und weiterzuentwickeln und diese anschließend in Rechenprogramme umzusetzen. Die Grundlagen und Ergebnisse seiner Simulationsrechnungen stellt er hier vor.

Bayreuth, im Dezember 2008   Professor Dr.-Ing. Dieter Brüggemann

Author's Prologue

This work was carried out at the Department of Engineering Thermodynamics and Transport Processes in the University of Bayreuth. I would like to thank Prof.Dr.-Ing. Dieter Brüggemann for giving me the chance and financial support to work in his staff and learn about such interesting and important topics for the environment and power generation. It has really been a great experience to explore and know more about the area of alternative energies.

Three full-length papers were published under his supervision:

- L. Matamoros and D. Brüggemann, Simulation of the water and heat management in proton exchange membrane fuel cells, Journal of Power Sources, 161, 203-213, 2006.

- L. Matamoros and D. Brüggemann, Numerical study on PEMFC's geometrical parameters under different humidifying conditions, Journal of Power Sources, 172, 253-264, 2007.

- L. Matamoros and D. Brüggemann, Concentration and ohmic losses in free-breathing PEMFC, Journal of Power Sources, 173, 367-374, 2007.

Such publications were achieved under a great demand of quality from the reviewers, given that "modeling and simulation of PEMFC" is a complex and saturated area. Hence, achieving to publish these papers about numerical modeling and simulation means a considerable progress of this work.

I would also like to thank Prof. Brüggemann's whole staff (Dr. Obermeier, Gabriele, Sabine, Christian, Andreas, Ernst, Stephan's, Manuel, Mark, Peter, Phang, Uli's, Markus) for being such good partners during this whole time.

Finally, I would like to thank my parents for all the unconditional and restless support during my whole life.

Kurzfassung

Unter den verschiedenen Typen von Brennstoffzellen ist die Polymerelektrolyt-(Protonenaustauschmembran-)Brennstoffzelle (PEMFC) besonders interessant, um einmal konventionelle Energiewandler zu ersetzen. Ihr Anwendungsbereich ist breit und umfasst die Stromversorgung für kleine elektronische Geräte, Fahrzeuge und Kraftwerke. Ein Wasser- und Wärmemanagement ist erforderlich, um hohe Leistungen in der PEMFC zu erreichen. Einlassströme werden gewöhnlich befeuchtet, um Ohmsche Verluste zu reduzieren. Jedoch kann das durch die Reaktion an der Kathode gebildete Wasser kondensieren, im porösen Material Poren verstopfen und den Transport der Reaktanden zu aktiven Zentren behindern. Die Temperatur fördert nicht nur die Reaktion an der Kathode und erhöht den Sättigungsdruck, sondern trocknet auch die Elektrolytmembran aus. Daher sind Wasser- und Wärmemanagement miteinander verknüpft zu betrachten. Hierfür sind die verschiedenen Transport- und elektrochemischen Vorgänge, die die PEMFC-Leistung beeinflussen, zu identifizieren und zu beschreiben. Modellierung und Simulation sind besonders nützliche Methoden, um PEMFC unter verschiedenen Bedingungen zu untersuchen und optimieren.

In der vorliegenden Arbeit wurde ein nicht-isothermes und drei-dimensionales numerisches Modell einer PEMFC entwickelt, um das Wasser- und Wärmemanagement unter verschiedenen Bedingungen zu berechnen. Das Hauptziel war, treibende und begrenzende Faktoren unter verschiedenen Feuchtebedingungen zu ermitteln und für verschiedene geometrische Dimensionen einzusetzen, um die Funktion der PEMFC bei hohen Leistungen zu vereinfachen. Die betrachteten Größen waren Geschwindigkeit, Temperatur, Massenbruch, Stromdichte, Spannungsverlust, Wassergehalt der Polymermembran, Sättigung und Flüssigkeitsströmungsfelder.

Ergebnisse zeigten, dass es abhängig von Design und Wassermanagement zu starken Begrenzungen des Stofftransports kommen kann. Die Leistung der PEMFC wird durch Austrocknen stark beeinträchtigt, kann jedoch durch Wahl geeigneter geometrischer Parameter beträchtlich verbessert werden. Es erweist sich als wichtig, gleichzeitig Ab- und Desorption des Wassers durch das Polymerelektrolyt und die Flüssigkeitssättigung im porösen Material der Kathode und Anode zu betrachten. Darüber hinaus wurden frei atmende PEMFC untersucht, um Konzentrationsverluste unter eingeschränkten Transportbedingungen der Reaktanden quantitativ zu bestimmen. Zusammengefasst sollten PEMFC mit mäßiger Befeuchtung zum Einsatz kommen, um höhere Gesamtleistungen zu erzielen und Konzentrationsverluste zu vermeiden, die durch Sättigung mit Wasser bei hohen Stromdichten entstehen.

Abstract

Among the different types of fuel cells, the proton exchange membrane fuel cell (PEMFC) turns up as the most attractive candidate to substitute traditional energy converters. Its range of application is wide and comprises power supply to small electronic devices, transportation and power generation plants. Water and heat management are necessary to achieve high power outputs in PEMFC. Inlet flows are usually prehumidified to reduce ohmic losses caused by non-optimal hydration of the polymer electrolyte. However, the water produced by the cathode reaction and the lack of diffusion for the disposal of this water may cause condensation in the cathode porous mediums that may clog pores and make difficult the reactants transport to active sites. Temperature not only stimulates the cathode reaction and increases the saturation pressure but dehydrates the polymer electrolyte as well. Hence water and heat management are closely related and must be properly established. It is necessary to identify and understand several transport and electrochemical processes that affect the PEMFC power output. Modeling and simulation are the most usefully, fast and inexpensive tools for study and optimization of PEMFC under different conditions.

In present work, a non-isothermal and three-dimensional numerical model of a PEMFC was developed to compute the water and heat management under different conditions. The main objective was to determine driving and limiting factors under different hydrating conditions and using different geometrical dimensions, in order to simplify PEMFC operation keeping high performances. Output variables were velocity, temperature, mass fraction, current density, voltage loss, water content of the polymer membrane, saturation and liquid flow fields.

Results showed that there may be severe mass transfer limitations depending either on the design or on the water management of the cell. The performance of PEMFC is seriously affected under dehydrating conditions. However, such performance may be considerably improved by using suitable geometrical parameters. These results show the importance of simultaneously calculating both the water absorption and desorption through the polymer electrolyte and the liquid saturation in the cathode and anode porous mediums to obtain an actual view of ohmic and concentration losses of the PEMFC performance. Free breathing PEMFC was also studied in order to determine concentration losses under such limited reactants transport conditions. Numerical model and simulations proved to be suitable to study diverse phenomenon that limit PEMFC performance, and so important conclusions about PEMFC operation were accomplished. In the whole, PEMFC should be able to operate under moderate hydrating conditions, in order to achieve higher overall efficiencies and avoid concentration losses produced by water saturation at high current densities. PEMFC should reduce as much as possible its complex operation (fuel, water and heat management), in order to become more attractive for commercialization.

Contents

List of Figures

List of Tables

Nomenclature

Symbol	Denomination	Unit
a	water activity	
ARH	anode relative humidity	
CRH	ambient relative humidity	
A_{superf}	Pt area per Pt mass	m^2kg^{-1}
A_{LG}/V_L	gas-liquid interfacial area	m^{-1}
C	molar concentration	$kmolm^{-3}$
C_{O_2s}	oxygen molar concentration at agglomerate surface	$kmolm^{-3}$
C_{O_2ref}	oxygen reference molar concentration	$kmolm^{-3}$
c_p	volumetric thermal capacity	$Jkg^{-1}K^{-1}$
CCL	cathode catalyst layer	
CCLT	cathode catalyst layer thickness	m
CGDLT	cathode gas diffusion layer thickness	m
D	diffusivity	m^2s^{-1}
E	cell potential	V
F	Faraday constant	
G	Gibb's free energy	J
GCW	gas channel width	m
H	height	m
ΔH	stored chemical energy	(Kgm^2s^{-2})
H_r	enthalpy of reaction	$(Jkg^{-1}K^{-1})$
i_{oref}	reference exchange current density	Am^{-2}
k	conductivity	$Wm^{-1}K^{-1}$
K_p	hydraulic permeability	m^2
L	length	m
m_{pt}	Pt mass per volume	kgm^{-3}
M	molecular weight	$kgkmol^{-1}$
N	molar flux	$kmolm^{-2}s^{-1}$
n	charge exchange	
N_{Drag}	electro-osmotic drag factor	$kmol_{H_2O}(kmol_{H+})^{-1}$
$\dot{n}_{GLH_2O}$	volumetric condensation rate	$kgs^{-1}m^{-3}$
P	pressure	Pa
PM	polymer membrane	
PMT	polymer membrane thickness	m
Q	heat flux	W
R_{agg}	agglomerate ratio	m
r	reaction rate	$kmols^{-1}m^{-3}$
RH	relative humidity	

Continuation

Sf	reduced saturation factor	
sf	saturation factor	
S	entropy	JK^{-1}
ST_ϕ	source term	
T	temperature	K
V	velocity	ms^{-1}
W	work	$Kgm^2s^{-2})$
x	molar fraction	
y	mass fraction	

Greek

α_c	transfer coefficient cathode side	
β	coefficient of volumetric expansion	K^{-1}
ϵ	porosity	
φ	efficiency	
η_c	cathode overpotential	V
δ	thickness	m
Γ	general diffusive coefficient	
ϕ	Property	
θ	contact angle	rad
λ	water content	$kmol_{H_2O}(kmol_{SO_3^-})^{-1}$
μ	dynamic viscosity	$kgm^{-1}s^{-1}$
ϱ	density	kgm^{-3}
σ	surface tension	Nm^{-1}
υ	stoichiometric coefficient	

Subscripts

agg	agglomerate
aq	aqueous
amb	ambient
c	capillary
ci	critical of specie i
CL	catalyst layer
GC	gas channel
GDL	gas diffusion layer
G	gas
L	liquid
V	momentum
T	energy
H_2O	water
O_2	oxygen

Continuation

H^+	proton
e^-	electron
SO_3^-	sulfonate group
im	immobile
i	species i
j	species j
m	mixture

Chapter 1

INTRODUCTION

1.1 Historical importance of alternative energy sources and converters

The combustion has been the most important motor for the revolutionary development of worldwide economy in last century, being the electrical generation and internal combustion motors its main applications. It basically consists of the oxidation of hydrocarbons; mostly producing carbon dioxide, water and heat.

Hydrocarbons used in combustion applications have been directly obtained either from oil or natural gas which are fossil non-renewable resources in the underground. Even though the availability of such resources is limited, its consumption has considerably increased worldwide, given its high feasibility, simplicity of extraction, transport and transformation. The world population growth and new emergent economies have also played an important role in this high world non-renewable energy demand. The result of such worldwide energy hunger has been the non-natural exhaust of tons of carbon dioxide and other combustion gases to the atmosphere. Such contaminant (or non-natural) emissions together with deforestation are responsible for a considerable "greenhouse" effect in the atmosphere, which is believed to either cause or accelerate the "global warming" according to many scientists worldwide.

Together with environmental aspects, the considerable increase of worldwide energy demand has stimulated the costs of fossil non-renewable energy sources. In a scenario of fossil fuel gradual depletion and increasing demand, such prices might get even much higher than today's ones. According to diverse estimations, oil production may reach its maximum in 5 years. Natural gas, which is assumed to substitute oil volumes to satisfy world fuel demand in the future, may reach its production peak in 50 years (there is not a certain estimation). Coal, which boosted industrial revolution in 18th and 19th centuries, may supply the whole world energy demand for 60 years given its high proven reserves. Nevertheless, stabilization of greenhouse gases is of much more importance than ups and downs of fossil fuel production. Therefore changes have to be made, in order to introduce alternative energy systems that substitute fossil non-renewable energy sources. New energy sources and converters have to be efficient, cheap, intense and clean to compete against the traditional ones. Although the efficiency of internal combustion with fossil fuels is quite low (<30%), the feasibility of this process is quite high and so its substitution is not easy to perform. New energy sources are being used keeping the scheme of internal combustion (e.g. biofuels). Without considering the efficiency of such new fuels, their environmental suitability is the main motivation for their use, given that they do not contribute to the increase of carbon dioxide emissions to the atmosphere.

Nuclear, Aeolian and Solar energy are also some examples of stationary substitutes for traditional fossil energy sources. In case of mobile applications, fuel cells have been regarded as one of the first candidates to substitute internal combustion motors given their high efficiency and clean emissions. Nevertheless, fuel cell feasibility and application are relative low for different reasons and so present work is aimed for the study of these systems in order to add contributions for the improvement of their performance.

1.2 Electrochemical mechanisms of PEMFC

Fuel cell technology beginnings can be found over 160 years back in time, due to Sir William Grove. However, its development has been interrupted by the interest and use of other more feasible technologies. Main disadvantages of fuel cell have not only been its cost, but its operation as well, given the diverse complications to obtain high and efficient performances from its electrochemical mechanisms. Nevertheless, its high thermodynamic efficiency and clean emissions make it very attractive to be used as both stationary and mobile energy source, despite the aforementioned disadvantages. For these reasons, increasing efforts have been made to reduce costs and improve performance, in order to successfully introduce fuel cells to the mass commercialization. Basically, developments of new materials and optimization have been the focus of researchers to obtain higher performances and lower costs.

A fuel cell is generically defined as a device that can convert the chemical energy of a reaction directly into electrical energy [1]. Among the different types of fuel cells, the proton exchange membrane fuel cells (PEMFC) turns up as the most attractive candidate for the substitution of internal combustion motors, given its low temperature of operation $\approx$ 80 °C, high efficiency and power density. Nevertheless, its main advantages (low temperature and high power density) may also be its main disadvantages, given that water is likely to condense at low temperatures and high power output. The performance of PEMFC may be highly affected as a product of the clogging of void spaces and gas channels which affects the reactant transport to the active sites where the reactions take place. Consequences are limited performance and so higher active surfaces.

A PEMFC comprises an anode and a cathode. Both are separated by a polymeric electrolytic membrane. Fuel and oxidant inlet flows enter to the system by gas channels, from which reactant species diffuse through a porous layer (gas diffusion layer), that enhances the uniform distribution of reactants to the active sites for reaction (catalyst layer). These active sites are located among these porous layers and the polymeric membrane. The sub-system that is composed by gas diffusion layers, catalyst layers and polymer membrane is called "MEA" (Membrane and Electrode Assembly). Fig. 1.1 shows a general scheme of the PEMFC components. Fuel is gaseous hy-

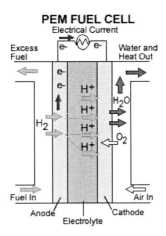

Figure 1.1: Basic scheme of PEMFC adapted from www.eere.energy.gov

drogen, which is separated into protons on platinum catalyst particles supported by carbon black porous pellets in the anode catalyst layer:

$$H_2 \rightarrow 2H^+ + 2e^- \qquad (1.1)$$

Protons flow through the polymer membrane to the cathode catalyst layer to complete the reaction on others active sites with oxygen (provided by an air flow):

$$\frac{1}{2}O_2 + 2H^+ + 2e^- \rightarrow H_2O \qquad (1.2)$$

Water, heat and electrical current are the final products. The electrical current is given by the donation of electrons made by hydrogen. The donated electrons travel to the cathode by an external circuit, from which electrical work is obtained. Under high current exchange (approx. 1 Acm^{-2}), efficiency of this process could reach 60%, which is a considerable improvement in comparison to the internal combustion efficiency.

High obtainable efficiencies and clean emissions under low temperatures are the main advantages of the electrochemical process described above. Nevertheless, there are still some steps remaining for the optimization and boost of PEMFC technology for mass commercialization.

1.3 Brief description of standard PEMFC components

PEMFC systems comprise several configurations and applications like serpentines, interdigitated, free-breathing, etc. PEMFC's components can be generically described as follows:

1.3.1 Polymer membrane

The polymeric membrane is composed by hydrated perfluorinated ionomeric materials [2]. This material consists of three regions: (1) polytetrafluoroethylene as backbone, (2) side chains of -O-CF2-CF-O-CF2-CF2- which connect the molecular backbone with, (3) ion clusters consisting of sulfonic acid ions. When the polymer electrolyte becomes hydrated, protons in the third region become mobile by bonding to water molecules and moving between sulfonic acid sites [3]. If the polymer electrolyte is well-hydrated, the resistance to the jumping of protons between sulfonic acid sites is relative small. Additionally, the protons drag water molecules when jumping between sulfonic sites. Such phenomenon is called electro-osmotic drag; the amount of water molecules dragged by one proton depends on the amount of water content of the polymer. Fig. 1.2 shows the molecular structure of Nafion which is usually the polymer electrolyte used in PEMFC.

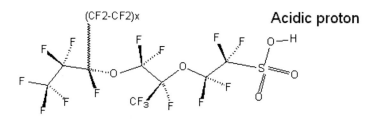

Figure 1.2: Polymer molecular structure (Nafion)

1.3.2 Catalyst layers

Catalyst layers consist of polymer electrolyte mixed together with carbon black particles. These particles bear Platinum particles, on which reactions take place. Fig. 1.3 shows an ideal representation of the catalyst layer structure.

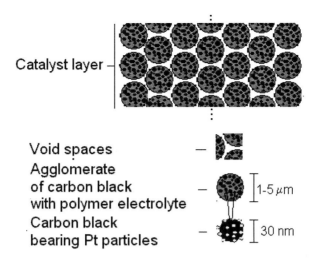

Catalyst layer –

Void spaces –

Agglomerate
of carbon black – $1\text{-}5\,\mu m$
with polymer electrolyte

Carbon black – $30\,nm$
bearing Pt particles

Figure 1.3: Representation of PEMFC's catalyst layer

1.3.3 Gas diffusion layers

Gas diffusion layers (Fig. 1.4 from [4]) consist usually of carbon fiber paper, which is coated with Polytetrafluoroethylene (Teflon) to stimulate the dispose of water that may condense in its porous structure.

Figure 1.4: Photo of PEMFC's gas diffusion layer (Nam and Kaviany, 2003)

1.3.4 Bipolar plates and gas channels

Gas channels are carved in the bipolar plates to allow the axial flow of reactants. The bipolar plates not only distribute the reactants and products, but also carry the electrical current through the gas channel ribs in and out of the MEA and facilitate the heat exchange and disposal of the condensed water. Fig. 1.5 shows the "standard" representation of a unit of a PEMFC stack which comprises the components mentioned above.

Finally, fuel cell stacks work like series of batteries. Each fuel cell unit (Fig 1.5) is

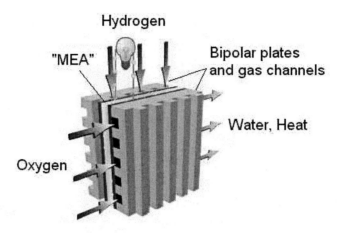

Figure 1.5: Scheme of PEMFC Stack adapted from www.eere.energy.gov

set up in series to achieve significant voltages.

1.4 Performance of PEMFC

Fuel cell performance highly depends on its operational conditions. Parameters like inlet flows compositions, temperature and pressure play the most important role when analyzing the power output of a fuel cell system. Additionally, the energy losses produced by internal resistances to the chemical reactions (cathode and anode overpotentials) and electrolyte transport may also influence the overall performance of a fuel cell system. Hence the optimization and understanding of these systems become a difficult task, given the amount of parameters involved in the whole electrochemical and transport process.

Next items describe the thermodynamical and chemical concepts involved in the determination and representation of fuel cell performance. In thermodynamical terms, every fuel cell is different, and as such next concepts only apply for PEMFC with platinum.

1.4.1 Cell polarization

Work produced by a fuel cell is given by the difference of free energy change of each half-electrochemical-reaction:

$$\Delta G = \Delta G_{cathode} - \Delta G_{anode} \tag{1.3}$$

6

Half-electrochemical-reaction free energy change can be expressed as the sum of reactants and products chemical potentials (μ_i):

$$\Delta G = \sum_{i=1}^{n} n_i \mu_i \tag{1.4}$$

For the anode reaction, the equation 1.4 can be expressed as:

$$\Delta G_{anode} = 2\mu(H^+(SO_3^-)) + 2\mu(e^-(Pt)) - \mu(H_2(g)) \tag{1.5}$$

And for the cathode reaction, the equation 1.4 can be defined as:

$$\Delta G_{cathode} = \mu(H_2O(g)) - \frac{1}{2}\mu(O_2(g)) - 2\mu(H^+(SO_3^-)) - 2\mu(e^-(Pt)) - \mu(H_2(g)) \tag{1.6}$$

Maximal work available of the fuel cell is achieved under reversible conditions ($\Delta S = 0$). Basically the electrochemical reaction rate should be 0 to obtain this potential maximal work (absence of irreversibilities). Maximal cell work can also be expressed in terms of maximal cell potential under reversible conditions, which is obtained from the difference between the half-electrochemical-reaction free energy change (eqs. 1.5 and 1.6) in absence of exchange current density (reversible conditions) as:

$$E^\circ = \frac{\Delta G^\circ}{nF} \tag{1.7}$$

For PEMFC using platinum as catalyst, the maximal reversible cell potential without exchanged current (E°) is 1.23 V [1, 5]. However, this value may change as a result of products and reactants activities different than unity. Non-standard temperatures ($\neq$ 298 K) may also linearly affect the reversible cell potential value. Therefore, the final form of this parameter can be expressed as follows (Nernst equation, see appendix A):

$$E = E^\circ + \frac{RT}{nF} \ln P_{H_2} P_{O_2}^{\frac{1}{2}} - 0.0009(T - 298) \tag{1.8}$$

For an irreversible process (real operation), the available work is less than the ideal as a result of voltage losses produced by ohmic losses and reaction resistances in the fuel cell system. Therefore, the equation 1.8 can be transformed for real operation as follows:

$$E = E^\circ + \frac{RT}{nF} \ln P_{H_2} P_{O_2}^{\frac{1}{2}} - 0.0009(T - 298) - \eta_o \tag{1.9}$$

where η_o represents the voltage losses caused by ohmic losses and reaction resistances, which increase with the intensity of exchanged current. Different phenomena may play important roles in the definition of these irreversible losses of PEMFC, which could be classified as:

- Activation losses or cathode and anode overpotential η_c and η_a, respectively (resistances to reaction rate at every electrode)

- Ohmic losses in the conductive phases (mainly produced by resistances to the proton flux in the electrolytic phase, the ohmic losses of the rest of the conductive phases are usually considered as negligible)

- Concentration losses (mainly produced by diffusional controls and barriers, reactant depletion and reactant cross-over between the electrodes).

Thus, the best way to express the performance of a fuel cell (PEMFC in this case) is by using polarization curves, which show the cell potential at different current densities. Doing so, it is possible to analyze the behavior of a fuel cell system under different conditions at low, moderate and high current densities. Fig. 1.6 shows an example of the utility of a polarization curve in the analysis of a fuel cell performance.

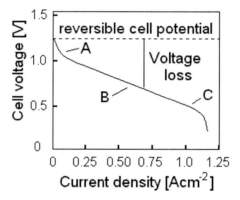

A: domain of activation polarization
B: domain of ohmic losses in polymer electrolyte
C: domain of concentration losses

Figure 1.6: Description of polarization curve

As shown in the Fig. 1.6, the loss of voltage for the activation of the electrochemical reaction governs the cell performance at low current densities. This voltage loss increases for higher current densities. From this point, ohmic losses in the conductive phases start being considerable and mostly dominate the cell performance for moderate and high current densities. The behavior of the polarization curve tends to be linear because the resistance to either electron or proton flux is theoretically constant. The resistance of the polymer electrolyte to the proton flux turns out to be one of the most important factors to consider in PEMFC performance at moderate and high current densities. Such resistance is defined by the energy cost of one proton to jump between sulfonate sites in the polymer electrolyte. This "jump" of protons may be considerably alleviated when the polymer electrolyte is hydrated, given that water molecules serve as transport medium when protons jump from one sulfonate site to another. Therefore, water content of polymer electrolyte is of paramount importance for the appropriate operation of these systems. PEMFC may not work when the polymer electrolyte is considerably dehydrated, and so an appropriate water balance must be kept inside the system to achieve suitable performances. At higher current densities, concentration losses play the most important role. Water condensation is usually the main factor that causes the power output drop at high current densities in PEMFC.

1.4.2 Cell efficiency

The efficiency of an energy converter φ can be defined as the relation between the amounts of produced useful energy (W) and stored chemical energy (ΔH) [1]

$$\varphi = \frac{W}{\Delta H} \tag{1.10}$$

The efficiency turns out to be considerably low for internal combustion engines, given the concept of Carnot cycle maximal efficiency which defines a maximal performance for devices operating between two "heat reservoirs" as

$$\varphi_{max} = \frac{Q_{in} - Q_{out}}{Q_{in}} = 1 - \frac{T_{cold}}{T_{hot}} \tag{1.11}$$

The Carnot cycle is driven by the difference of temperature between the "cold" and "hot" reservoirs. Therefore an energy converter defined by this process cannot operate without rejecting part of the input energy. Given the operational complexities, the real efficiency is usually rather low for several applications of the Carnot cycle.

Internal combustion is actually defined by the Otto cycle, which is an open cycle and driven by the compression of fuels and presence of flames in a cylinder with a piston. Given the mechanical characteristics of such devices, most of the heat produced by the reaction is released to the environment; only a relative small amount (compared to the fuel available energy) is transformed into mechanical work. Therefore the efficiency is usually rather low for such machines ($\leq 30\%$).

Fuel and oxidant are separated by an electrolyte in a fuel cell and the reaction between them is driven by the difference of free energy change at each half-cell-reaction which causes an electrical current from one electrode to another. Each reaction takes place at catalysts, and so the operational temperature could be considerably reduced. The oxidation of fuels is carried out by an electrochemical process, which can be controlled, and not by either heat transfer or combustion, and so the efficiency of such processes is considerably higher than any "standard" process defined by the Carnot cycle, basically due to the fact that thermal energy is mostly an output and not an input.

Following the eq. 1.10, the efficiency of PEMFC can be defined from eq. 1.8 and eq. 1.9 as

$$\varphi_{PEMFC} = \frac{E}{E° + \frac{RT}{nF}\ln\frac{P_{H_2}^2 P_{O_2}}{P_{H_2O}} - 9.10^{-4}(T - 298)} \tag{1.12}$$

Hence, the behavior of PEMFC efficiency can also be indirectly observed in Fig. 1.6. The reversible cell potential represents the maximal achievable work ($\varphi_{PEMFC} = 1$) and the cell potential expresses the work obtained from the fuel at real operation. Therefore, the shape of polarization curve is equal to the efficiency's one. Likewise, the highest efficiency can be achieved when the current density is near to 0. Then, the efficiency becomes lower for higher current densities as a result of the increase of different power losses formerly described.

High active areas would be needed to achieve high efficiencies at low current densities, which is not practical at all. Therefore, another concept must be introduced in addition to the polarization curve in order to enhance the analysis of a fuel cell system. The power density obtained from the product between cell potential and current density (P=E.I) is usually chosen to observe the maximal power of a fuel cell system under different current densities and conditions. Fig. 1.7 shows an example of the behavior of the power output of a PEMFC. Even though power density increases along with the current density, there is an inflexion point where ohmic and concentration losses become too high to achieve higher power output at high current densities, consequently the performance of the cell starts diminishing.

1.4.3 Main advantages and limiting aspects of PEMFC

In comparison to most of energy converters, the main advantages of PEMFC may be listed as:

Main advantages:

- High efficiency ($\approx 60\%$: more than double of standard internal combustion).

- Low temperature of operation: basically due to the use of noble metals as catalyst, allowing high performances at low temperatures (60-80 °C).

- Quick start-up: low operation temperatures reduce start-up time and so high performances can be achieved in short times.

- Clean emissions when operating with pure hydrogen: Maybe the most significant advantage given the future changes of ambient laws. As seen before PEMFC emission is water when using pure hydrogen as fuel and it turns out to be a strong point when considering ambient contamination as an aspect to reduce and avoid in the future.

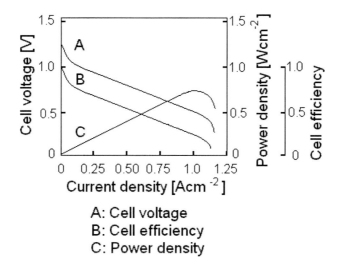

Figure 1.7: Efficiency, polarization and power density curves

and the main limiting aspects may be exposed as follows:

Main limiting aspects:

- Slow cathode kinetics: 1000 times slower than anode kinetic and so PEMFC performance is basically governed by cathode conditions. Therefore high active area is needed to achieve high output, and so costs considerably increase.

- Sensitivity to CO containing fuels: main disadvantage of noble catalysts as platinum. It considerably limits the use of alternative fuel, given that pure hydrogen industry is in development, and a bridge between traditional fuel and pure hydrogen is obligatory for the economical success of this technology. Therefore PEMFC systems using reformers and shift reactors to reduce CO fraction in reformed hydrocarbons to 100 ppm is imperative when analyzing a possible and near introduction of PEMFC to mass commercialization.

- Complex water and heat management: water and heat management should take many complex phenomenons into account in order to achieve optimal operation under different conditions.

- Fuel management: where does hydrogen come from? Hydrogen may be obtained from water. However such process is highly expensive. Use of light hydrocarbons like natural gas and methanol is very attractive. However the use of reformers and shifters to convert such fuels into hydrogen introduces new complications and so the PEMFC efficiency may be highly affected given that such process are achieved at 800 and 400 °C, respectively. Afterwards, the produced hydrogen must be cooled to around 80 °C for the electrochemical process. This whole reforming process considerably reduces the feasibility of PEMFC.

• High production and operation costs: expensive materials and high active areas considerably increase the total cost of PEMFC. However the costs in manufacture may be considerably reduced in conditions of mass production.

Among the limiting aspects listed above, water and heat management area is considered by this work as highly interesting and of high importance, given that performance of PEMFC strongly depends on the behavior of vapor and liquid water in the system. Thus next sections are focused on the important aspects of water and heat management.

1.4.4 Water and heat management

The inlet flows are usually injected as a mixture with water vapor at a certain temperature to avoid the polymer electrolyte dehydration and so alleviate the resistance of polymer electrolyte to proton flux. However, the water produced by the cathode reaction and the lack of diffusion for the disposal of this water may cause condensation in the cathode porous mediums that might clog pores and make difficult the transport of reactants to active sites. The gas diffusion layers are usually coated with polytetrafluoroethylene (PTFE) to enhance the disposal of condensed water to the gas channels and so alleviate the possible flooding.

The temperature not only stimulates the cathode reaction and increases the saturation pressure but dehydrates the polymer electrolyte as well. Therefore the temperature plays an important role in water management and so water and heat management are closely related.

Water and heat management of PEMFC are the most important aspects to account for, especially at high current densities, given that both polymer electrolyte dehydration and water condensation in porous mediums may cause severe damage to the performance of PEMFC. Therefore, the water and heat management should be properly established to avoid both dehydration of the polymer electrolyte and flooding in the cathode porous mediums. In other words, the concentration and ohmic losses must be diminished as much as possible, in order to achieve higher performances at high current densities under a wide range of conditions.

It is necessary to identify and understand several transport and electrochemical processes that affect the PEMFC power output, in order to enhance and optimize the operation of these devices under different conditions. Experimentation and numerical modeling have been made and utilized to analyze and optimize the water and heat management in PEMFC systems. Next sections show a brief description of the most significant experimental and numerical works of PEMFC water and heat management area.

1.5 Review of PEMFC experimentation

Experimental determination of local properties of PEMFC is highly difficult, due to the small geometry especially in the through-plane direction of gas channels and the MEA (membrane and electrode assembly) [6–8]. Only output variables like average voltage and current density may be easily measured. However the use of polarizations (voltage vs. current density) is insufficient when analyzing the PEMFC performance, given the amount of parameters involved in the electrochemical process. There are not only several transport parameters to be studied, but also many geometrical and electrochemical factors that may affect the function of PEMFC. Thus, experimental analysis on PEMFC become not only rather complicated, but time consuming as well. Putting all these facts together, modeling and simulation become the most usefully, fast, inexpensive and practical tool for study and optimization of PEMFC. Nevertheless, much experimental efforts should be aimed for the determination of important distributions like current density, in order not only to improve the understanding of physicochemical phenomena in PEMFC, but also to carry out better simulations for the design and development of PEMFC systems [9]. Up to now, some experimental setups [10–14] have been developed to get approaches of current density distributions in PEMFC, which are important advances to validate numerical data and improve the understanding of PEMFC phenomenon.

Experimentation is important to achieve the measurement of material properties needed to carry out the correct water and heat management in a PEMFC system. Important experimental works have been made to determine the significant parameters that govern the water transport in the polymer electrolyte, in order to quantitatively understand the behavior of the water content under several conditions. Many of these works have been considered as reference when making the input for modeling and simulation of PEMFC assuming Nafion as polymer electrolyte. For example, Springer et al. [2] presented experimental data for the determination of Nafion water content in equilibrium with the water activity of the surrounding gas at 30 °C. Zawodzinski et al. [15] studied the water content of membranes exposed to liquid water and water vapor at different activities. The equilibration of polymer electrolyte with liquid was different than the equilibration with saturated water vapor (Schroeder's paradox). Hinatsu et al. [16] presented experimental data of the equilibrium water content against water activity at 80 °C. Broka and Ekdunge [17] observed from experimental tests that water content decreases dramatically in range of temperature of 25-50 °C and change is slower for higher temperatures. However, Dannenberg et al. [18] suggested making a linear interpolation to determine water content for temperatures in 30-80 °C range, given that the variation is almost linear.

Springer et al. [2], Fuller and Newman [19] and Zawodzinski et al. [15] presented experimental data concerning the electro-osmotic drag of water by proton flux in poly-

13

mer membrane at 30 °C. Neubrand [20] presented experimental data for the water electro-osmotic drag at 80 °C. Concerning the water diffusion coefficient of water in polymer electrolyte, Zawodzinski et al. [15] experimentally determined values at 30 °C and used a correction factor for different temperatures. Neubrand [20] presented a correlation to estimate this parameter at 80 °C.

When using Nafion as polymer electrolyte, its water content is the most important factor to account for in PEMFC operation. Hence, the determination of flow, compositions and temperature fields under several conditions is needed to understand factors that may affect water transport in the whole system. Consequently, polymer electrolyte characterization is an important task to correctly and numerically determine the water and heat management. In this work Nafion is the polymer electrolyte by default.

1.6 Review of PEMFC numerical modeling

Many researchers have been focused on making mathematical models to account for transport behaviors of PEMFC. These models have been created for calculation of one, two and three dimensional variations of important properties inside the fuel cell. However, they have made emphasis on different aspects. In the 90's, Springer et al. [2] presented an isothermal, one dimensional, steady state model for calculation of transport through the electrodes. As aforementioned, they experimentally determined important parameters involved in water transport through polymer electrolyte (Nafion). Bernardi and Verbrugge [21] also presented an 1-dimensional model to determine mass fractions profiles and saturation of water. However, they considered the polymer membrane as fully hydrated. Fuller and Newman [19] made a two dimensional model which accounts for thermal effects and water management in polymer membrane. Gurau et al. [22] published a two dimensional model, which describes transport of momentum, energy and mass in polymer membrane, catalyst layers, gas diffusion layers and gas channels by using Darcy's law together with Navier-Stoke's equations. Yi and Nguyen [23, 24] provided a two dimensional model to account for hydrodynamics and multicomponent transport in interdigitated flow fields. These models provide information about the importance of water and heat management in PEMFC, however their applicability is limited.

Water management becomes highly important for high performance PEMFC, and so there should be more complex models to study behavior of species movement. Um et al. [25] presented a transient multidimensional model for electrochemical kinetics, current distribution, fuel and oxidant flow, and multicomponent transport. CFD formulation in electrochemical systems [26] was extended to PEMFC systems for the first time by these authors [25]. Even though this model represents a more complex solution in comparison to earlier one and two dimensional models, ohmic losses in polymer electrolyte under dehydrating conditions and water condensation were neglected. Dutta et al. [27, 28] presented a numerical model for calculation of three dimensional variations of velocities and components mass fraction using CFD approach. Membrane electrode assembly (MEA) was not included in simulations, consequently water transport and ohmic losses in polymer electrolyte were not properly taken into account. Liquid water saturation in porous mediums was omitted.

Liquid water condensation is an important factor to consider especially at high current densities. Presence of condensed water in porous mediums can not be overseen. Clogging and flooding in cathode may considerably affect PEMFC performance at high current densities. Therefore two-phase modeling in PEMFC represents a difficult and obliged step forwards in comparison to former works. New models have been published to calculate liquid saturation in PEMFC porous mediums. Two-phase modeling in PEMFC porous mediums may be divided into: multiphase mixture approach (M2) [26] and separate phases approach (unsaturated flow theory UFT belongs to this category) [7, 29]. Wang et al. [30] presented an isothermal two-phase flow and multicomponent transport model based on multiphase mixture approach [26]. M2 consists of making averages from phases properties to obtain a mixture property. Flow is first assumed and calculated as a mixture. Afterwards liquid water field is obtained from taking into account capillary flow from bulk flow. Natarajan and Van Nguyen [6, 31] presented a two-phase model which assumes only-diffusion-driven-flux in cathode porous medium, and so M2 is omitted. Liquid water flow is driven by capillary forces which is similar to the former approach of [26, 30]. However, Wang et al. [30] used Leverett's function to calculate capillary pressure and cubic factors to account for effect of saturation on permeability. Natarajan and Van Nguyen [6, 31] used a mathematical function to calculate capillary pressure, and permeability was linearly related to saturation. They mentioned the origin of Leverett's function which is a result of correlating experimental data from porous mediums like soil. Natarajan and Van Nguyen's correlation is based on fitting model results to PEMFC experimental data. You and Liu [32] presented a work highly similar to [26, 30] based on M2. Berning et al. [7, 29] proposed a liquid water velocity field driven by pressure and capillary pressure gradients. Both gas and liquid flow fields at cathode porous medium are modeled by using Darcy's law, and so both phases are assumed to flow separately. Nam and Kaviany [4] proposed a model describing movement of water in gas diffusion layer by assuming only diffusion mechanisms in cathode porous mediums, and so liquid flow was only driven by capillary forces. Leverett's function is used in a different presentation, as former works [29, 30], to estimate capillary flow for hydrophobic porous mediums. Pasaogullari and Wang [33, 34] proposed a model to determine liquid water transport based on [30] for hydrophobic porous mediums. They also made a comparison between UFT and M2 approaches. They concluded that M2 is more suitable to model two-phase phenomenon involved in cathode porous mediums.

The aforementioned models basically depend on water activity, interfacial area and permeability. Nevertheless, M2 models may yield an incorrect view of two-phase flow in porous mediums, given that phases flow separately under low volumetric flows. Only dispersed flow patterns under turbulent convection like dispersed bubble flow and mist flow may agree with this assumption. Convection forces in PEMFC cathode porous medium may be negligible given the high resistance to flow and then capillary forces dominate liquid flow in cathode porous medium.

A boundary layer of liquid saturation is needed to determine interfacial GDL/gas channel capillary flow when assuming unsaturated flow theory (UFT) driven by capillary forces. Wang [9] mentioned that interfacial saturation should not be equal to immobile saturation [4] due to the dependence of this factor on diameter of liquid droplets and population density. Meng and Wang [35] expressed the interfacial saturation as dependent on gas velocity, contact angle and current density. Zhang et al. [36] experimentally showed that interfacial saturation depends strongly on gas velocity in gas channels. Two flow patterns in gas channel were recognized: mist and annular flow (with occurrence of slug flow). Each flow pattern is product of balances between drag forces and interfacial saturation. A proper estimation of liquid interfacial saturation and liquid water behavior would be interesting to improve the performance of PEMFC simulations.

The effects of the parameters, like gas channel and ribs width as well as the porosity of the gas diffusion layers, on the PEMFC performance were recently studied by Guvelioglu and Stenger [37] and Lum and McGuirk [38] using CFD modeling. In these works, isothermal operation and single phase flow were assumed. The interfacial polymer membrane water content was evaluated using Springer et al. [2] equilibrium correlation at 30 °C. This means that the diffusion resistance in the non-equilibrium process of water transfer through the polymer membrane interface was not considered. The boundary condition at the polymer membrane interface should be modeled by diffusion in both mediums when using differential modeling. Otherwise the water transfer in and out of the polymer membrane can not be established. It is important to mention that the equilibrium polymer electrolyte water content at higher temperatures than 30 °C should be estimated, given that the polymer electrolyte is dehydrated by increasing the temperature. The data of Hinatsu et al. [16] at 80 °C can be useful to achieve an estimation of the equilibrium water content of the polymer electrolyte at temperatures between 30 and 80 °C by linear interpolation. When assuming catalyst layers as a surface, ohmic losses in this region are considered as negligible. It is important to model the resistance to the proton flux exerted by the polymer electrolyte contained in the cathode catalyst layer, especially at dehydrating conditions, given that such process may affect the activation potential of the catalyst, and so the sensitivity of the numerical modeling under different conditions may be different.

1.7 Motivations and objectives

It has been established that use of numerical tools is highly important to predict behaviors and distributions that are experimentally difficult to obtain under diverse conditions. In order to achieve a complete solution of the transport and electrochemical phenomena in PEMFC, the water management, which in turn requires the consideration of the heat management, must be appropriately determined. Hence, the influence of water on PEMFC performance should be properly simulated, and as such the water movement in polymer membrane as well as the condensation and capillary flow in the cathode gas diffusion layer and catalyst layer should be properly estimated. Finally, the flow field in the gas channel should be able to dispose the condensed water. If these aspects are neglected, the numerical results may be erroneous and misleading, given that the behavior of the whole system is highly dependant on liquid and gaseous water transport. Therefore, modeling and numerical simulation of PEMFC should be aimed for simultaneously calculating the water transfer through the polymer membrane and catalyst layers interface as well as the liquid water saturation in porous mediums. PEMFC's simulations under different hydrating conditions become essential when numerically analyzing the water and heat management. Estimations of concentration and ohmic losses of power under different conditions are important for the optimization of PEMFC. Realistic numerical results over a wide range of conditions should be obtained, in order to numerically determine the most limiting factors in PEMFC operation and how they could be diminished. Understanding and predicting of power losses in the PEMFC control volume are imperative to propose solutions and strategies.

As seen before, many numerical models have been created to account for water and heat management [4, 6, 7, 25, 27, 29, 30, 34, 35, 37–40]. The modeling aspects considered as highly important by present work can be classified as follows:

- As aforementioned, the diffusion resistance in the non-equilibrium process of water transfer through the polymer membrane interface should be considered. The boundary condition at the polymer membrane interface should be modeled by diffusion in both mediums when using differential modeling. Otherwise the water transfer in and out of the polymer membrane can not be established, and so the water balance in the whole cell may be erroneous. All of the works cited before do not account for any actual water balance in the whole PEMFC system. The interfacial water content of the polymer membrane is calculated assuming equilibrium with the bulk gas phase, which does not consider the species and whole mass balance in the system. Such condition can not establish an actual water exchange between the polymer membrane and the surroundings.

- The equilibrium polymer electrolyte water content at temperatures higher than 30 °C should be estimated, given the dehydration of polymer electrolyte at higher temperatures. The data of Hinatsu et al. [16] at 80 °C can be useful to achieve an estimation of the equilibrium water content of the polymer electrolyte at temperatures between 30 and 80 °C by linear interpolation. Most of the works cited before does not account of the dehydration of polymer electrolyte at temperatures higher than 30 °C, which may cause misleading results.

- When assuming catalyst layers as a surface, ohmic losses in this region are considered as negligible. It is important to model the resistance to the proton flux exerted by the polymer electrolyte contained in the cathode catalyst layer, especially at dehydrating conditions, given that such process may affect the activation potential of the catalyst, and so the sensitivity of the numerical modeling under different conditions may be different. This aspect is usually ignored and the catalyst layers are treated as active flat surfaces without transport resistances.

- Even though concentration losses in anode side are considered insignificant in the present work, liquid water saturation in the anode porous mediums should be included in simulations, given that this phenomenon may influence water transfer through the polymer membrane and anode porous medium interface. In other words, presence of liquid water in anode porous mediums affects the anode water balance, and so a fraction of water may be either absorbed or desorbed by the polymer membrane, another portion may condense and the rest may be disposed by the gas flow in channel. Consequently overall ohmic losses may be affected by both cathode and anode condensation of water, which means that this phenomenon should be taken into account, even though concentration losses in anode are considered as negligible.

The main objective of this work is to present a numerical solution of the water and heat management in PEMFC. The polymer membrane is considered as a component highly affected by its external conditions, hence the mass transfer between the membrane, the cathode and the anode porous mediums is taken into account, in order to determine the actual water content field, and so the actual voltage losses in polymer membrane and the cathode electrolyte overpotential are obtained. An electrochemical and phenomenological model is proposed and based on several important phenomena that must not be taken for granted when studying water and heat management in PEMFC. Effects and behaviors observed from present work are of high importance and interest.

Influences that some geometrical parameters (thickness of polymer membrane, cathode catalyst layer and gas channel to rib width ratio) may have on the water transport and as such on the performance of a straight PEMFC under different hydrating conditions are analyzed, in order to study the diverse phenomena that may dominate electrochemical mechanisms in these systems and improve the performance and utility of PEMFC modeling and simulation. Important conclusions concerning enhancement of PEMFC performance were achieved.

Considering the motivations and objectives formerly exposed, this work can be defined as a numerical solution for the study and analysis of PEMFC components. Behaviors of PEMFC under diverse conditions are observed. Actual driving and limiting factors are pointed out which can be useful for the improvement of the research and development of this technology. An advanced conclusion of this work is the suitability of the operation under dehydrating conditions when using appropriate geometrical parameters. Ohmic losses can be considerably reduced and operation under dehydrating conditions is likely to yield high performances.

The validation of data is an obliged step of every either experimental or numerical work. Considering that each author has made emphasis on different aspects of the PEMFC, the comparison between numerical models becomes useless. It is necessary to point out that the success of the validation depends on the use of the correct electrochemical input data and the correct polymer electrolyte characteristics. Additionally, the geometrical parameters should be as close as possible to the parameters of the experimental setup used as reference. Comparisons between experimental and numerical distributions become of high importance not only to confirm the suitability of the numerical data, but also to improve the understanding of different important phenomenon that may take place in the control volume. Up to now, some experimental setups [10–14] have been developed to get approaches of current density distributions in PEMFC, which are important advances, given that the polarization curves may be misleading for data validation [9]. In this work, some of data of Ihonen et al. 2001 [41], Mench et al. 2003 [42] and Ju et al. 2005 [43] is used to show the suitability of this model to study the behaviors of PEMFC over a wide range of current densities and conditions. The numerical input data was adjusted to the experimental conditions of each data source, in order to achieve a reliable validation.

Chapter 2

METHODOLOGY AND MODELING

As aforementioned, this work is focused on achieving a numerical solution of the PEMFC water and heat management in order to perform simulations under diverse conditions for the understanding and optimization of PEMFC power output. Data validation is achieved from comparing numerical results to polarization and current densities distributions from the literature [41–43]. Such procedure was considered sufficient to regard the numerical solution as an useful tool to study and predict PEMFC performance. The methodology of this work can be described as follows.

2.1 Fundamental equations and source terms

The model is three-dimensional and simultaneously calculates the variables of gas channels, catalyst layers and polymer membrane in a straight PEMFC. The most limiting effects of PEMFC power occur at the through-plane direction. Therefore the straight PEMFC is the most simple, fast and suitable configuration to analyze the phenomena that could limit the PEMFC performance. Other geometries may be more suitable to study the in-plane behavior, which would be important to analyze other aspects like heat transfer in and out of the stack. However, the transport in the through-plane direction highly dominates the performance of PEMFC and so in-plane phenomena can be considered as of second importance when studying the most significant limiting aspects of these systems. Hence, large-scale simulations are not an objective of present work and as such this study in focused on the control volume seen in Fig. 2.1. Using a symmetry assumption, the domain shown in Fig. 2.1 was vertically split in two parts and only the left part was calculated.

Conservation balances and source terms are showed in Table 2.1. Diffusion is assumed to be the primary mechanism of transport in the gas diffusion layers and catalyst layers while the convection is assumed as negligible. The pressure drop in gas channels (source term of momentum in gas channels) is estimated using the Darcy friction factor, consequently the use of coupled pressure-velocity equations was avoided. The source term of the continuity equation can be ignored [44].

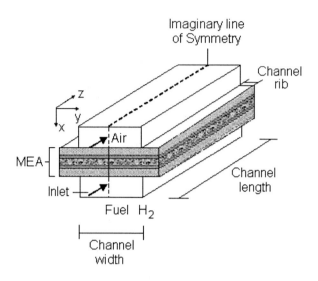

Figure 2.1: Control volume of the straight PEMFC for the numerical solution.

Joule heating in polymer membrane, heat produced in cathode reaction and heat produced in condensation-evaporation entropy change are taken into account as source terms of energy ST_T. Joule heating ST_{T1} in the membrane is calculated by using the proton conductivity in Nafion k_{H+} and the proton flow. It is important to mention that the energy balance in the polymer membrane was calculated assuming the thermal conductivity as 100 $Wm^{-1}K^{-1}$ [22]. Heat produced in the cathode reaction is determined using the reaction heat of oxygen reduction ΔHr_{O_2} and reaction rate, which basically consists of heat produced by the entropy change and cathode overpotential [5]. Condensation-evaporation entropy change in cathode porous mediums is estimated using the latent heat of vaporization of water at local temperature H_{LG} and the volumetric condensation rate.

Table 2.1: Conservation balances and source terms

Balance	Differential expression	Source terms
Momentum	$V \cdot \nabla \varrho V = -\nabla P + \nabla \mu \nabla V + ST_V$	$ST_V = 0$
Energy	$V \cdot \nabla \varrho c_p T = -\nabla k \nabla T + ST_T$	$ST_{T1} = \frac{(N_{H+}F)^2}{k_{H+}}$
		$ST_{T2} = \Delta Hr_{O_2} r_{O_2}$
		$ST_{T3} = H_{LG} \dot{n}_{GLH_2O}$
Species	$V \cdot \nabla \varrho y_i = -\nabla D_{i,m} \nabla \varrho y_i + ST_i$	$ST_i = M_i r_i$
		$ST_{H_2O} = \dot{n}_{GLH_2O}$
Continuity	$\nabla \varrho V = ST_m$	$ST_m = 0$
Polymer membrane	$\nabla D_\lambda \nabla \lambda = \frac{M_{SO_3^-}}{\varrho_{SO_3^-}} \frac{dN_{Drag}N_{H+}}{dx} + ST_\lambda$	$ST_\lambda = 0$
Liquid balance	$\nabla \frac{\varrho_{H_2O(L)} K_p Sf^3}{\mu_{H_2O(L)}} \left(\frac{dP_c}{dSf}\right) \nabla Sf = ST_{sf}$	$ST_{sf} = \dot{n}_{GLH_2O}$

The electrochemical kinetic was determined by using the agglomerate model [8, 17, 45–47]. The electrochemical reaction rate obtained from the agglomerate model is considered as source term of species ST_i in catalyst layers. The stoichiometrical factor of every species is derived from the well-known mechanism of reaction in PEMFC. The agglomerate model states that the catalyst layer is a porous medium composed by platinum and carbon particles mixed together with polymer electrolyte. This mixture is defined as spherical agglomerates, which have void spaces between them (Fig. 1.3). In this work, the polymer electrolyte is assumed to be as Nafion, the water is assumed to be produced in gas phase and condensation could take place in this medium. The whole electrochemical reaction is assumed to be dominated by the cathode reaction kinetic, and so the reaction rate is finally expressed as

$$r_{H+} = \frac{3D_{agg}C_{O_2s}(1-\epsilon)}{R_{agg}^2}(\theta R_{agg} \coth \theta R_{agg} - 1) \tag{2.1}$$

with the Thiele Modulus (θR_{agg}) defined as

$$\theta R_{agg} = \sqrt{\frac{m_{Pt}^v A_{superf} i_{oref}}{FD_{agg}C_{O_2ref}}} R_{agg} \exp\left(-\frac{\alpha_c F\eta}{2RT}\right) \tag{2.2}$$

The Henry constant for the dissolution of oxygen in water is used to calculate the concentration at the agglomerate surface. Bernardi and Verbrugge [21] presented an empirical correlation to calculate the Henry constant of oxygen in liquid water at different temperatures and pressures. Hence the oxygen concentration in Nafion is estimated from the Henry constant of oxygen in water as follows (see appendix B.1 for a more detailed description of Bernardi and Verbrugge work [21]):

$$C_{O_2s} = x_{O_2}(P/atm) \exp\left(\frac{666}{T/K} - 14.1\right) \tag{2.3}$$

The reference exchange current density (i_{oref}) is calculated by the correlation presented by Wang et al. [8] based on the experimental data of [48] at different cell temperatures:

$$\log(i_{oref}) = 7.507 - \frac{4001}{T/K} \tag{2.4}$$

The diffusivity of oxygen in Nafion D_{agg} is calculated from the correlation presented by Marr and Li [46], which is derived from a curve fit published in [48] (see appendix B.2 for a more detailed description of Parthasarathy et al. [48] work):

$$D_{agg} = D_{O_2-Nafion}\epsilon_{agg}^{1.5} = \left[-1.07.10^{-5} + 9.02.10^{-6}\exp\left(\frac{T-273}{106.7}\right)\right]\epsilon_{agg}^{1.5} \tag{2.5}$$

The factor $\epsilon_{agg}^{1.5}$ represents the Bruggemann relation used to adjust the diffusivity of oxygen in Nafion to the catalyst layer porous medium.

It is important to mention that the parameter m_{Pt}^v represents the volumetric cathode catalyst loading which consists of the relation between the cathode catalyst loading per area and the cathode catalyst layer thickness as follows:

$$m_{Pt}^{v} = \frac{m_{Pt}}{\delta_{CCL}} \qquad (2.6)$$

The product of $m_{Pt}^{v} A_{superf}$ yields the volumetric platinum area, which is coupled with the reference exchange current density on platinum surface i_{oref} to define the current output of the cell without considering the concentration and ohmic losses. The cathode catalyst loading value of $m_{Pt} = 1.0$ mgPtcm^{-2} was arbitrarily chosen as base case, and its performance depends on the surface area value ($A_{superf} = 250$ cm^{2}Pt/mgPt), which was also arbitrarily chosen in the range of $110 - 1400$ cm^{2}Pt/mgPt according to Marr and Li data [46]. Therefore, the performance of present PEMFC directly depends on these factors that numerically yield moderate and high current densities under diverse conditions.

To account for the voltage losses induced by the Nafion resistance to the transport of protons to active sites, an 1D-expression of Ohm's law is defined to estimate the increase of overpotential in the cathode catalyst layer:

$$k_{H^+}[(1 - \epsilon)\epsilon_{agg}]^{1.5}\frac{\delta\eta}{\delta x} = N_{H^+}F \qquad (2.7)$$

where $k_{H^+}[\Omega^{-1}m^{-1}]$ is the proton conductivity in Nafion expressed as [2]

$$k_{H^+} = (0.5139\lambda - 0.3260)\exp\left[1268\left(\frac{1}{303} - \frac{1}{T}\right)\right] \qquad (2.8)$$

This expression is modified by using the Bruggemann factor $[(1 - \epsilon)\epsilon_{agg}]^{1.5}$ to account for the catalyst layer and agglomerate porosity. This expression (eq. 2.8) shows the strong dependence of the proton conductivity of Nafion on the water content and can be applied for $\lambda \geq 1$ (see appendix B.3 for a more detailed description of Springer et al. [2] work).

Ohmic losses in anode catalyst layer are omitted because of the considerable faster hydrogen oxidation kinetics in comparison to oxygen reduction's, and as such anode overpotential is assumed to be negligible. This assumption may only be affected by catalyst poisoning when using CO containing hydrogen flows. However, catalyst poisoning is also omitted in this work. In present work, anode catalyst layer could be seen as a portion of polymer membrane in electrochemical terms.

Two-phase flow in the cathode porous medium (gas diffusion layer and catalyst layer) is assumed as separated phases. Berning and Djilali [29] proposed a liquid water velocity field considering both convection and capillary forces. In this work the liquid is assumed to flow individually driven by capillary pressure gradients, produced by gradients of liquid saturation and surface tension [4, 6], consequently the convection is considered as negligible in porous mediums. The capillary pressure is semi-empirically calculated by means of the Leverett function J(Sf) for hydrophobic unsaturated porous mediums [4, 34]:

$$P_c = \frac{\sigma \cos \theta}{(K_p/\epsilon)^{\frac{1}{2}}} J(Sf) \tag{2.9}$$

The Leverett function J(Sf) for hydrophobic porous mediums is expressed as

$$J(Sf) = 1.417Sf - 2.120Sf^2 + 1.263Sf^3 \tag{2.10}$$

where the reduced saturation factor Sf is the following

$$Sf = \frac{sf - sf_{im}}{1 - sf_{im}} \tag{2.11}$$

Below the saturation value sf_{im} (immobile saturation), the water is assumed to be a discontinuous flow pattern, and as such the capillary flow can be assumed as negligible. At the gas channel, the liquid water is disposed either by high convective air flow or by evaporation. Then, the reduced saturation factor is assumed to be zero at the gas channel and gas diffusion layer interface [4].

In order to calculate the water change of phase to establish the saturation factor in porous medium, the condensation kinetic theory is used to determine the volumetric condensation rate $\dot{n}_{GLH_2O}$ which is expressed as

$$\dot{n}_{GLH_2O} = K_{GL} M_{H_2O} \frac{A_{LG}}{V_L} \frac{P_{H_2O(G)} - P_{H_2O}^{Sat}}{RT} \tag{2.12}$$

The factor K_{GL} consists of diffusion terms for the condensation process [4]. The evaporation was assumed to be the inverse of the condensation process. The gas-liquid interfacial area A_{LG}/V_L was arbitrarily assumed as constant. It is important to mention that the condensation and evaporation process in porous mediums is considered as additional source term for water gas balance (species balance in table 2.1).

To calculate fuel cell potential, the cathode overpotential and polymer membrane voltage loss are subtracted from the equilibrium potential. The voltage loss produced by bipolar plates and contact resistance between gas channel ribs and gas diffusion layers was assumed as negligible:

$$E = 1.23 - 0.0009(T - 298) + \frac{RT}{2F} \ln(P_{H_2} P_{O_2}^{\frac{1}{2}}) - \eta_c - \frac{i_{avg} \delta_{PM}}{k_{H_+}} \tag{2.13}$$

2.2 Transport Properties

Gases are assumed as ideal. Dynamic viscosity is determined using the kinetic theory of gases.

In order to consider the influence of porous medium on the conductive transport, the following correction factors (Eq. 2.14 and Eq. 2.15) were used for catalyst layers and gas diffusion layers, respectively [49]:

$$f(\epsilon) = \epsilon^{1.5} \tag{2.14}$$

$$f(\epsilon) = \epsilon \left(\frac{\epsilon - \epsilon_\beta}{1 - \epsilon_\beta} \right)^\alpha \tag{2.15}$$

The factor ϵ_β is a percolation threshold considered as 0.11 and α is an empirical constant considered as 0.527 for in-plane diffusion and 0.785 for cross-plane diffusion [4].

Thermal conductivities (k) are expressed as an average between conductivities of phases involved in each infinitesimal element of the system

$$k = k_S f(1 - \epsilon) + k_G f(\epsilon)(1 - sf) + k_L f(\epsilon)sf \tag{2.16}$$

Thermal capacity (c_p) is also regarded as an average between phases. Consequently, thermal equilibrium between phases is assumed:

$$c_p = c_{pS}(1 - \epsilon) + c_{pG}\epsilon(1 - sf) + c_{pL}\epsilon sf \tag{2.17}$$

Mixture diffusivity was approximated by the mixture rule [50]:

$$D_{i,m} = \frac{\sum_{i=1, i \neq j}^n x_i M_i}{\sum_{i=1}^n x_i M_i \sum_{i=1, i \neq j}^n \frac{x_i}{D_{i,j}}} \tag{2.18}$$

Binary diffusivities were calculated using the following correlation for gases at low density [50]:

$$\frac{PD_{i,j}}{(P_{ci}P_{cj})^{\frac{1}{3}}(T_{ci}T_{cj})^{\frac{5}{12}}(\frac{1}{M_i} + \frac{1}{M_j})^{\frac{1}{2}}} = a \left(\frac{T}{\sqrt{T_{ci}T_{cj}}} \right)^b \tag{2.19}$$

with a and b constant. If the mixture diffusivity is calculated for porous mediums, Eq. 2.14 and Eq. 2.15 are used to account for the effect of porosity and tortuosity of the catalyst layer and gas diffusion layer, respectively. The effect of liquid water saturation is accounted for by using a normalized function $f(sf) = (1 - sf)^{1.5}$. Therefore, the final form of the diffusivity is

$$D_{i,m-\text{corrected}} = D_{i,m}f(\epsilon)f(sf) \tag{2.20}$$

For the polymer electrolyte, the electro-osmotic drag coefficient N_{Drag} was calculated at 30 °C and 80 °C using the experimental data for Nafion from Zawodzinski et al. [15] and Neubrand [20], respectively. From both empirical data [15, 20], the electro-osmotic drag was adjusted to the following expression for polymer membrane water content ranging from 0 to 22 (see also [9]):

$$N_{Drag} = 0.1125\lambda \tag{2.21}$$

Concerning the diffusion coefficient of water in the polymer electrolyte, the following approach (Eq. 2.22) was made from the experimental data of Zawodzinski et al. [15] (30 °C) (see appendix B.4 for a more detailed description of Zawodzinski et al. [15] work) and Neubrand [20] (80 °C) to determine the value of this property.

$$D_\lambda = 1.5.10^{-7} \lambda e^{-\frac{2436}{T}} \tag{2.22}$$

2.3 Boundary conditions

In order to determine the interfacial water transfer between the cathode and anode porous mediums and the polymer membrane, the difference between the bulk concentrations of water in each phase and the equilibrium concentrations of water at the interface was used as driving force of absorption and desorption of water by the polymer electrolyte. The convective mass transfer is neglected, and so the water exchange is expressed in terms of the molar fraction of each phase (Y_i and X_i) and the mixture diffusivity of species i by each phase ($D_{i,m1}$ and $D_{i,m2}$):

$$\varrho_1 D_{i,m1} \frac{X_i - X_i^*}{M_1 \Delta x_1} = \varrho_2 D_{i,m2} \frac{Y_i^* - Y_i}{M_2 \Delta x_2} \tag{2.23}$$

The equilibrium data from Springer et al. [2] at 30 °C and from Hinatsu et al. [16] at 80 °C were used to complete this interfacial mass transfer model by relating the molar fraction of water of each phase at the interface (Y_i^* and X_i^*). The interfacial polymer membrane water content at different temperatures is calculated by linear interpolation.

$$
\begin{aligned}
\lambda_{30C} &= 0.043 + 17.81a - 39.85a^2 + 36a^3 & &\text{for} \quad a \leq 1 \\
\lambda_{80C} &= 0.3 + 10.8a - 16a^2 + 14.1a^3 & &\text{for} \quad a \leq 1 \\
\lambda_{30C} &= -9.1776 + 37.276a - 15.877a^2 + 2.265a^3 & &\text{for} \quad 1 < a < 2.5 \\
\lambda_{80C} &= -30.412 + 61.978a - 25.96a^2 + 3.7008a^3 & &\text{for} \quad 1 < a < 2.5 \\
\lambda &= 20 & &\text{for} \quad a \geq 2.5
\end{aligned} \tag{2.24}
$$

Liquid uptake is ignored in this work. Modeling liquid water uptake may be a difficult task, given the complexities inherent to molecular structure of polymer electrolyte and liquid water distribution in gas diffusion layers and catalyst layers. Zawodzinski et al. [15] showed complexities inherent to water uptake from vapor and liquid water (Schroeder's paradox). However liquid water uptake may be direct and easier than water vapor uptake, given the minimization of surface barrier to liquid water [15]. Equilibrium water content of polymer electrolyte (λ) immersed in liquid water is around 22 at 30 °C according to Zawodzinski et al. [15] data. Nevertheless, water uptake from vapor phase is likely to mostly hydrate the polymer membrane [15], because that catalyst layer and gas diffusion layer must be almost flooded to account for a significant liquid water uptake.

Considering that ohmic losses are also calculated in the cathode catalyst layer, the water content of the polymer electrolyte contained in this region is assumed to be uniform and equal to the calculated interfacial water content of polymer membrane.

2.4 Numerical solution

All equations of balance were discretisized and solved simultaneously using the Jacobi-Gauss-Seidel relaxation iterative numerical method. This method and the boundary conditions were programmed using Fortran90 for calculation of modeling output parameters. Output data was processed using Matlab. Even though the whole code, input and output data configuration represented a considerable effort, its presentation is omitted in this work (a scheme of the program structure and discretization is presented in appendix C).

Use of CFD commercial software was dismissed, given the complexity of the iterative calculation of the interfacial water content of the polymer membrane through the equation 2.23 which is achieved using the method of bisection in a system composed by two equations and two dependant variables (actually one equation and one variable). If using commercial software, the diffusive resistance for the interfacial water transfer through the polymer membrane interface would have been assumed as negligible and so the water balance in the system would have been erroneous. In other words, water balance in polymer membrane would have been independent of water balance in porous mediums, and as such water fraction in the whole system may have been overestimated. As mentioned before, the amount of water in the system is respected, and so an appropriate boundary condition between catalyst layers and polymer membrane should be established to account for the amount of water that is transferred in and out of the membrane. Considering that the performance of PEMFC highly depends on the amount and transport of water in the system, such procedure was considered as suitable to achieve realistic results under a wide range of conditions. The objective was "efficacy", putting emphasis on the dominant factors and phenomenon of PEMFC, in order to accomplish a numerical tool able to predict and show realistic results under several conditions for analyzing and optimizing these systems.

Chapter 3

RESULTS AND DISCUSSION

Once the numerical solution was ready for simulation, an input strategy had to be managed to achieve a set of numerical results interesting for analysis of PEMFC under several conditions. First important aim was the simulation of water and heat management under different operational conditions, in order to understand the behavior of PEMFC limiting phenomena. Different conclusions were achieved under this scheme. Afterwards, a set of simulations was carried out using diverse geometrical parameters under different humidifying conditions. Interesting conclusions were accomplished especially concerning reduction of ohmic losses. Free-breathing PEMFC was also included in this input data strategy, and so important results were obtain using this configuration useful for some special applications (like portable batteries). In the whole, enhancement of PEMFC can only be achieved by reducing the limiting factors that affect these systems in the through-plane direction. Better electrodes and more flexible polymer electrolytes would considerably improve the attractiveness of these devices. Even though a base Nafion-Platinum straight PEMFC is used as control volume in this work, numerical results are highly interesting for research and development of this technology.

3.1 Data validation

The validation of the data is an obliged step of every either experimental or numerical work. Considering that each author has made emphasis on different aspects of the PEMFC, the comparison between numerical models becomes useless. In this work, the numerical input data was adjusted as close as possible to the conditions of the experimental data from [41], in order to validate the numerical results over a wide range of current densities. The comparison between the experimental polarization curve at 80 °C and 200 kPa and the numerical one (Fig. 3.1) shows that the numerical solution is suitable for studying PEMFC systems.

Nonetheless, data validation is an important aspect of every work concerning numerical simulation. Therefore comparisons between experimental and numerical dis-

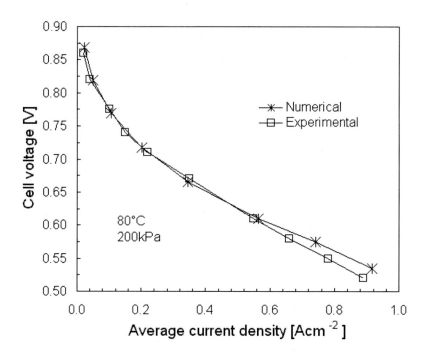

Figure 3.1: Experimental (Ihonen et al. 2001) and numerical polarization curves

tributions become of high importance not only to confirm the suitability of the numerical data, but also to improve the understanding of different important phenomena that
may take place in the control volume. Up to now, some experimental setups [10–14]
have been developed to get approaches of current density distributions in PEMFC,
which are important advances considering that the polarization curves may be misleading for data validation [9]. Thus results in Fig. 3.1 may not be seen as a reliable
and convincing validation. In this work, some of Mench et al. 2003 data [42] is used to
show the suitability of this model to study the behavior of PEMFC over a wide range of
current densities. The numerical input data was adjusted to the experimental conditions of [42] in order to achieve a reliable validation (input data in appendix D.1). The
numerical current density distributions show a reasonable increase when lowering the
cell potential and working under mass transport limitations (Fig. 3.2). This behavior
is in agreement with the experimental data presented in [42], consequently this modeling proves to be sensitive to different factors that may govern the performance of
PEMFC. At low potentials the shape of the current density distribution is dominated
by higher oxygen concentration at the entrance of the gas channels. The current density distributions show a decreasing trend caused by the oxygen depletion along the
gas channel. Experimental results do not show this behaviour at low potentials, but
an almost uniform distribution. The absolute difference between experimental and

numerical data (error) was 0.086V for the highest current density's case, which shows a reasonable behaviour of the numerical system when working at different potentials under fully humidifying conditions.

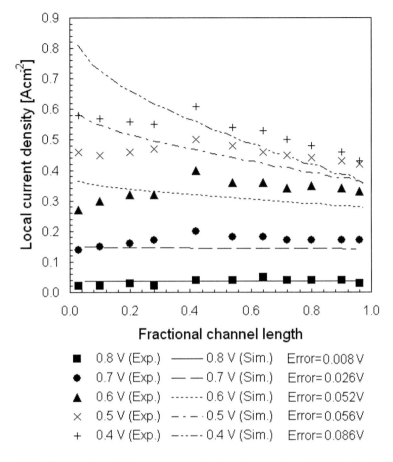

Figure 3.2: Comparison between numerical and experimental (Mench et al 2003) current density distributions under fully humidifying conditions

Data of Ju et al. [43] was used to make comparisons between numerical and experimental current density distributions under different humidifying conditions (input data in appendix D.2). Using the same geometrical, operational and electrochemical input data and even the modification of polymer membrane of this work (GORE-SELECT membrane or reinforced polymer membrane with half of water diffusion and proton conductivity of Nafion), results showed that numerical solution is in accordance to the realistic behavior of a PEMFC when operating at 0% cathode RH and 75% anode RH (Fig. 3.3). In this case, the deviation at the highest current density was 0.094V.

Likewise, performance of numerical solution was satisfying when comparing nu-

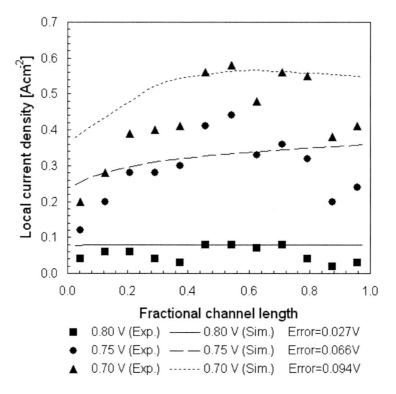

Figure 3.3: Comparison between numerical and experimental (Ju et al. 2005) current density distributions at 0% cathode RH and 75% anode RH

merical and experimental current density distributions at 50% cathode RH and 0% anode RH (Fig. 3.4). The highest difference between experimental and numerical data was 0.077V.

Former data validation showed the "suitability" of the numerical solution to study PEMFC systems. The numerical current density distributions were in accordance to the experimental measurements under different conditions, and so own modeling and simulation proved to be well-established, sensible and calibrated. The aim of this section was not to show the accuracy of the numerical solution but its suitability. Experimental setups also produce approaches given the different complexities inherent to the experimentation of PEMFC. In the whole, the trend of the numerical PEMFC was in agreement with experimental results. The difference between experimental and numerical data (defined as Error in the Figs. 3.2, 3.3, 3.4) is around the order of magnitude 0.1 for the highest current densities of every case, which means that the absolute error of the modeling and simulation in this work can be established in the range 10-20%.

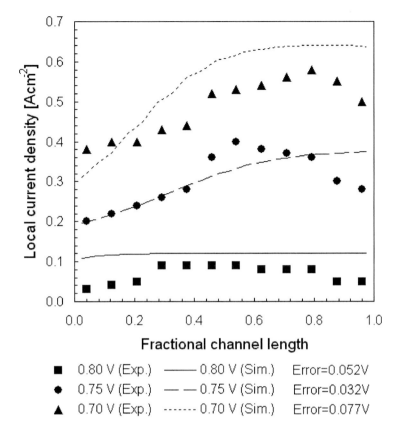

Figure 3.4: Comparison between numerical and experimental (Ju et al. 2005) current density distributions at 50% cathode RH and 0% anode RH

3.2 Simulation of the water and heat management of PEMFC

In PEMFC, there are many parameters that may affect their performance either slightly or heavily. These parameters may be separated into three sections: geometrical, electrochemical and operational. In order to perform simulations using different parameters, an appropriate base case must be chosen to obtain representative results. Geometrical, electrochemical and operational base parameters arbitrarily used in this work are listed as follows (table D.2).

The earlier aim of this study was to analyze the behavior of the polymer membrane water content, the saturation in cathode porous medium and the cathode reaction rate under different operational conditions, in order to establish the influence of these factors on the performance of PEMFC and prove the sensitivity of modeling under different conditions. The results are mostly presented as transversal averages along the gas channels, given that the three-dimensional displays are not practical to make

Table 3.1: Geometrical and electrochemical parameters

Description	Value
channel height	0.0004 (m)
channel width	0.001 (m)
channel length	0.10 (m)
channel rib	0.0002 (m)
gas diffusion layer thickness	0.0001 (m)
gas diffusion layer porosity	0.5
catalyst layer thickness	10 (μm)
catalyst layer porosity	0.4
polymer membrane thickness	100 (μm)
Pt loading	1 ($mg_{Pt}cm^{-2}$)
Pt surface area	250 (cm_{Pt}^2/mg_{Pt})
agglomerate porosity	0.2
agglomerate radii	0.00001 (cm)
Inlet Cathode Volumetric Flow	0.000001 (m^3s^{-1})
Inlet Anode Volumetric Flow	0.0000005 (m^3s^{-1})
Inlet Cathode Flow Temperature	353.0 (K)
Inlet Anode Flow Temperature	353.0 (K)
Inlet Cathode Flow Pressure	300 (kPa)
Inlet Anode Flow Pressure	300 (kPa)

comparisons.

3.2.1   Water content field at different current densities

The magnitude of the current density may affect the water content field of the polymer membrane. Therefore, the local current density and the local water content along the channel were computed at different total average current densities and fully humidified inlet flows (Fig. 3.5(A) and (B), respectively). In Fig. 3.5(A), it can be seen that the local current density diminishes along the channel, because of the depletion of oxygen by the cathode reaction. The local water content increases along the channel, given the increase of water activity, as a result of the production of water by the electrochemical reaction in the cathode (Fig. 3.5(B)). The reaction at the anode also contributes to the increase of the water activity at this side, due to the fully humidified condition of the anode inlet flow and the consumption of hydrogen along the channel. Consequently the water molar fraction increases along the channel at the anode side. The increase of water content becomes slower towards the end of the channel, as a product of the consumption of oxygen along the channel and so the reduction of the cathode reaction rate. Due to the fully humidified condition of the inlet flows, the electro-osmotic drag is insufficient to dry the anode side of the polymer membrane, even for high current densities. Under these conditions the cell voltage is only affected by the proton flux and not by the water content of the polymer membrane.

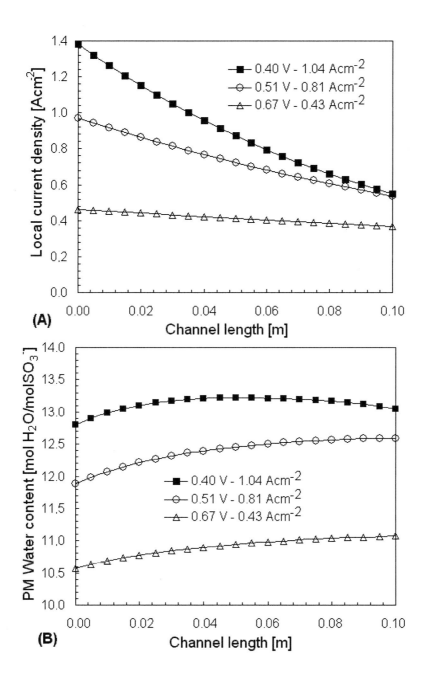

Figure 3.5: (A) Average current density and (B) polymer membrane water content along the channel for different average current densities

3.2.2 Effect of the cathode and anode inlet flow relative humidity

As mentioned above, when operating under fully humidifying conditions, the water content of the polymer membrane is favored, and consequently, the performance of the fuel cell is not affected by the hydrating conditions of the electrolyte. However, the relative humidity of the cathode inlet flow may have a considerably influence on the performance of the fuel cell. In case the inlet flows are not well humidified, drying of the polymer in the cathode as well as in the membrane will occur, causing an increase of voltage losses and a reduction of the cathode and anode reaction rate. If the water content decreases, cathode overpotential increases in active sites due to the increment of proton resistance. Thus cathode reaction rate may be considerably affected under such circumstances.

The relative humidity of cathode and anode inlet flows was varied at high current density operation, in order to study the performance of the fuel cell under dehydrating conditions. The relative humidity of the cathode inlet flow is maintained at 30% and the anode inlet flow relative humidity is set to 50, 70 and 90%. In this case, cathode and anode catalyst layer thickness was assumed to be equal to 1.0 μm. The Fig. 3.6(A) shows that the local current density diminishes along the channel, as a result of the decrease of oxygen concentration. The effect of the electrolyte on the current density is slightly seen for the anode relative humidity of 90%. The current density is higher at the gas channel inlet. However, the oxygen is faster consumed, and so, the local current density is less in comparison to the 50 and 70% relative humidity. The local water content increases along the channel (Fig. 3.6(B)), consequently the electrolyte limitations at the cathode catalyst layer decrease as well. In this case, the anode relative humidity enhances the hydration of the polymer membrane. Under these conditions it can be seen that the effect of the polymer electrolyte on the current density is small in comparison to the effect of the oxygen mass transfer limitations and depletion.

To study the potential effect of Nafion in the cathode electrode on the reaction, the thickness of the cathode catalyst layer was made thicker (by a factor of 10), maintaining the superficial platinum load constant, in order to increase the proton transport resistance in comparison to the reaction rate. The results are showed in Fig. 3.7(A) and (B) at the same humidity conditions used for the Fig. 3.6. The Fig. 3.7(A) shows the current density along the channel for dehydrating conditions. It can be seen that for low water content of polymer, the resistance of the Nafion to the proton transport becomes high and affect the performance of the fuel cell such as the oxygen mass transfer limitations do. In this case, the current density slightly increases along the channel, because the water content of the electrolyte becomes higher as a result of the production of water by the cathode reaction (Fig. 3.7(B)). Then, the oxygen mass transfer limitation and the decrease of the oxygen concentration reduce the reaction rate, despite the hydration of the polymer electrolyte.

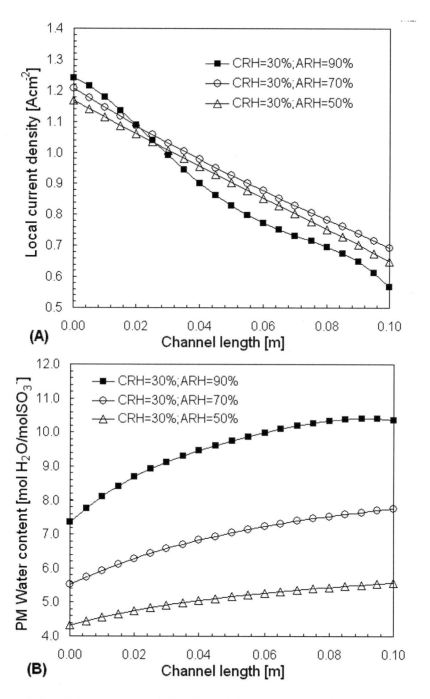

Figure 3.6: (A) Average current density and (B) polymer membrane water content along the channel for different relative humidities of anode inlet flow calculated for a catalyst layer thickness of 1.0 μm

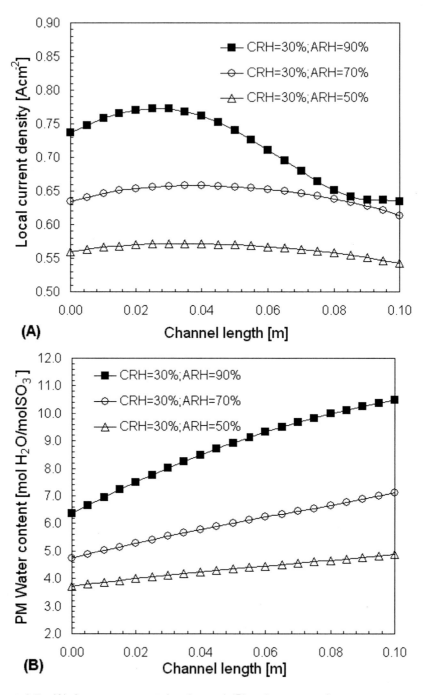

Figure 3.7: (A) Average current density and (B) polymer membrane water content along the channel for different relative humidities of anode inlet flow for a 10x thicker catalyst layer (10 μm) as compared to Fig. 3.6

The Fig. 3.8 shows the cathode water vapor mass fraction 3D-field, the polymer membrane water content 3D-field and the current density 2D-contour for relative humidity of cathode and anode inlet flows at 30 and 50% (respectively). As it can be seen, the maximal current density area has moved towards the end of the channel, as a result of the better hydration of the electrolyte by the electrochemical reaction. However, the current density is diminished, because of the decrease of the oxygen concentration along the channel.

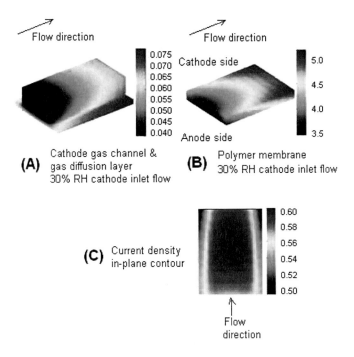

Figure 3.8: Maps of (A) vapor mass fraction, (B) polymer membrane water content field and (C) current density at 30 and 50% RH in cathode and anode inlet flows, respectively

3.2.3 Water saturation of the cathode porous medium

The cathode saturation represents one of the most important problems to consider, given that the liquid water may block the path of reactants to the active sites. Therefore, the inlet relative humidity improves the electrolyte performance of the fuel cell, but may decrease the flux for the diffusion to active sites. The saturation should be quantitatively estimated to avoid the flooding in the cathode side by an incorrect water management.

The saturation in the cathode porous medium may be affected by several factors:

current density, temperature of operation, hydrophobicity and permeability of porous medium and relative humidity of cathode and anode inlet flows. The condensation and evaporation process depends highly on the interfacial area between the gas and the liquid. To predict an accurate value for the volumetric interfacial area is a difficult task, especially for the complexity of the three-phase system. As mentioned above, the volumetric interfacial area is arbitrarily considered as a constant parameter during the condensation and evaporation process.

The saturation of the porous medium becomes a problem when the system is unable to dispose the liquid water product from the excess of water vapor at the same speed it is produced by the cathode reaction. Therefore, the liquid water accumulates and the available area for the diffusion is reduced. Then, high concentration losses occur, because of the high mass transfer limitations from the high mechanical resistance for diffusion. However, once the water accumulates, some continuous liquid water paths will be formed and there will be a capillary flow straight to the gas channel, so that the liquid water can be disposed. This phenomenon may keep the system from flooding, although the mass transfer control may remain.

In this work, the influence of the volumetric interfacial area on the performance of the fuel cell was computed by assuming different values of this parameter. The Fig. 3.9(A) and (B) show the local saturation and water content along the channel for volumetric interfacial areas of 1000, 10000 and 50000 m^2/m^3 at high current densities and fully humidified inlet flows, respectively. It can be observed from Fig. 3.9(A) that the amount of liquid water is slightly governed by the effect of the interfacial area, for the values chosen in this work. According to these results, the system is able to dispose the water and the maximal saturation is approximately below the value of 0.2 at the cathode catalyst layer. Consequently, the capillary flow controls the saturation process against the condensation and evaporation process. In this case the dominating effect is neither the volumetric interfacial area nor the current density of the fuel cell, but the permeability and hydrophobicity of the porous medium. It can also be seen from Fig. 3.9(A) how the average saturation changes along the channel. This factor increases as a result of the increase of water activity along the channel. The saturation tends to reach a maximum, given the decrease of reaction rate towards the channel. The water content could be affected by the volumetric interfacial area of the gas and liquid interface, as it can be seen in Fig. 3.9(B). The local water content decreases with increasing volumetric interfacial area. Therefore, the higher condensation of water vapor reduces the mass of water that may be transferred to the polymer electrolyte. Consequently, the voltage loss in the polymer membrane may be affected by the higher volumetric interfacial area.

The Fig. 3.10 shows the cell voltage and power vs. average current density for volumetric interfacial areas of 1000, 10000 and 50000 m^2/m^3. As it can be seen for high current densities, fuel cell voltage decreases while volumetric interfacial area

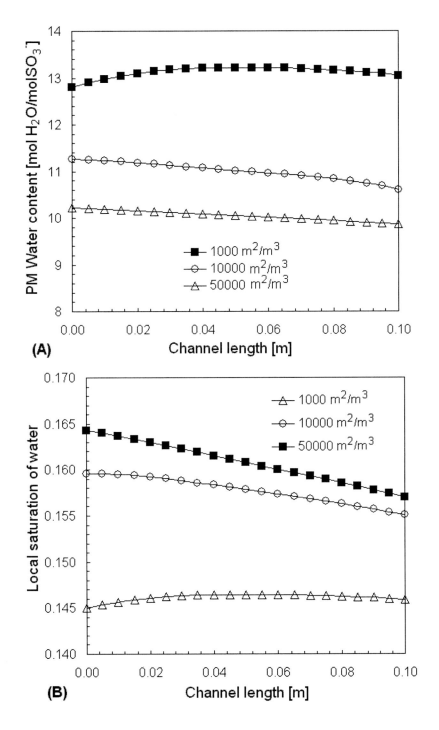

Figure 3.9: (A) Average polymer membrane water content and (B) cathode porous medium saturation along the channel at different volumetric gas-liquid interfacial areas

increases. Therefore, under these conditions the current density is not seriously affected by saturation of liquid water in the cathode electrode, given that the difference of the saturation magnitudes is small for all cases. The significance of this behavior may be the low mass transfer control that liquid water induces in the cathode porous medium in comparison to the resistance that the porous medium exerts on the diffusive flux. The performance of the fuel cell is mainly decreased by the influence of the volumetric interfacial area on the water uptake of the polymer membrane, and so on voltage of the fuel cell. Nevertheless, the water uptake from liquid water is neglected and this factor might be important to obtain more accurate results for the water uptake of the polymer membrane.

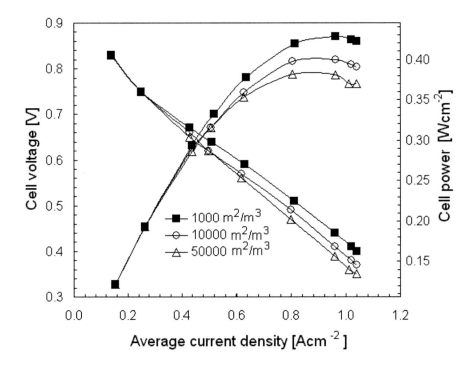

Figure 3.10: Cell voltage and power vs. average current density at different volumetric gas-liquid interfacial areas

To study the influence of the cathode inlet relative humidity on the liquid saturation of the cathode porous medium, different curves were computed at 30, 60 and 90% cathode relative humidity maintaining the anode relative humidity at 100% and the volumetric interfacial area at 1000 m^2/m^3 (Fig. 3.11). In this case the relative humidity has a strong influence on the condensation and evaporation process, due to the partial hydration of inlet flow. Then, there will be a tendency to evaporation at the gas channel and cathode gas diffusion layer, and so the water is disposed by change of phase and not by the drag of liquid droplet in the gas channel. For higher inlet humidity the capillary flow dominates against the evaporation, and so the disposal is made by the mechanical energy of the flow in the gas channel.

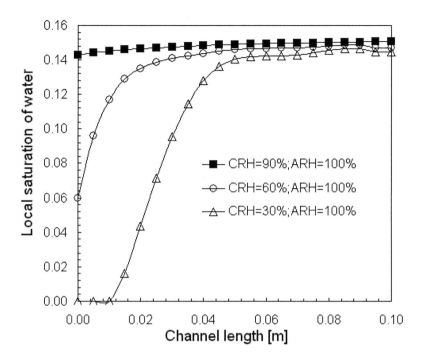

Figure 3.11: Cathode porous medium saturation along the channel at different cathode inlet flow relative humidity

The relation of local temperature and liquid saturation of the cathode porous medium is also to be accounted for. The Fig. 3.12 shows the distributions of temperature and liquid saturation in the through-plane direction along the channel right in the middle of the fuel cell for fully humidified inlet flows and high current density operation (1.04 Acm^{-2}). For fully humidified inlet flows, reaction rate is higher at the entrance and diminishes along the channel by depletion of reactants. Likewise temperature is higher at the entrance and becomes lower along the channel (Fig. 3.12(A)). In through-plane direction, temperature diminishes towards the gas channel due to convection in the interface between gas channel and gas diffusion layer. According to the distribution of temperature, the convection in gas channels maintains the cell almost at uniform temperature. Under the conditions listed in table. D.2, the heat of reaction and Joule heating are not high enough to increase considerably the temperature against convection in gas channels. In the case of the local saturation, the distribution (Fig. 3.12(B)) depends not only on temperature distribution, but also on water vapor mass fraction distribution and capillary flow. Fig. 3.12(B) shows that the local liquid saturation slightly increases along the channel and slightly decreases in through-plane direction towards the gas channel. Thus, as mentioned above capillary flow plays the most important role on the liquid distribution. Even though water activity may change almost 100% along the channel (as it can be seen in Fig. 3.8(A)), local saturation is almost homogeneous in cathode porous medium due to the capillary flow. The lower liquid saturation after the entrance of fuel cell shows that temperature and water activity distributions have an influence on the distribution of saturation; however these effects are small in comparison to the capillary flow.

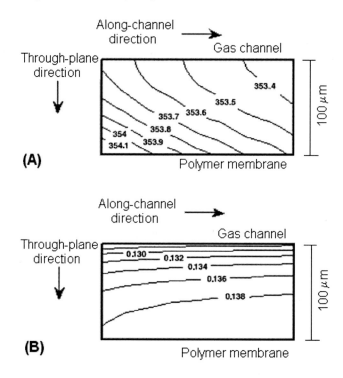

Figure 3.12: Through-plane and along-channel distributions of (A) temperature and (B) liquid saturation in the cathode porous medium

3.2.4 Effect of the operational temperature

To observe the influence of the operational temperature on the performance of the fuel cell, the local water content and the current density along the channel were computed for 70, 80 and 90 °C (Fig. 3.13(A) and (B), respectively). From Fig. 3.13(A), it can be seen that the local current density decreases while the temperature increases. This effect is especially seen for the results of 90 °C. Therefore, the performance of the fuel cell diminishes at higher temperatures for fully humidified inlet flows, as a result of the influence of the temperature on the hydration of the polymer electrolyte, as it can be seen in Fig. 3.13(B). For the temperature of 90 °C, the water content of the polymer membrane dramatically diminishes for fully humidified inlet flows, and so the electrolyte overpotential increases. Hence, the temperature may have a positive effect on the reduction of the oxygen, but the higher proton resistance produced by the electrolyte dehydration at high temperatures has a dominant influence on the cathode reduction reaction.

3.2.5   Effect of cathode inlet volumetric flow

At high current densities, the demand of oxygen is high, and so the inlet flow should be able to supply the needed amount of oxygen. In the case the oxygen is gradually depleted, the performance of the fuel cell diminishes, due to the high cathode overpotential induced by the lack of oxygen. The Fig. 3.14 shows the local current densities along the channel at different cathode inlet volumetric flows. The results show the high influence of the oxygen concentration on the cathode reaction rate. Even for the highest inlet volumetric flow, the drop of the current density along the channel is noticeable along the channel. Nevertheless, the channel length was arbitrarily chosen and the aim of this calculation is to show how the depletion of oxygen may affect the performance of the fuel cell at different current densities.

3.2.6   Partial conclusions I

In this whole set of results (from section "Simulation of the water and heat management of PEMFC"), the system showed to be highly dependant on the hydration of the polymer membrane, which was affected by the relative humidity of the inlet flows. The saturation of the cathode porous medium was dominated by the capillary flow, and so, the permeability was high enough to dispose the water in the system. The volumetric interfacial area of the water affected the performance of the fuel cell by reducing the amount of water to be transferred to the polymer membrane. The operational temperature may have a negative effect on the hydrating properties of the polymer electrolyte, and so on the performance of the fuel cell, especially for values above 90 °C. The saturation of the cathode porous medium was not dramatically affected by the operation conditions and its magnitude did not exceed the value of 0.2. Then, the capillary flow was high enough to dispose the produced water and to prevent the system from flooding.

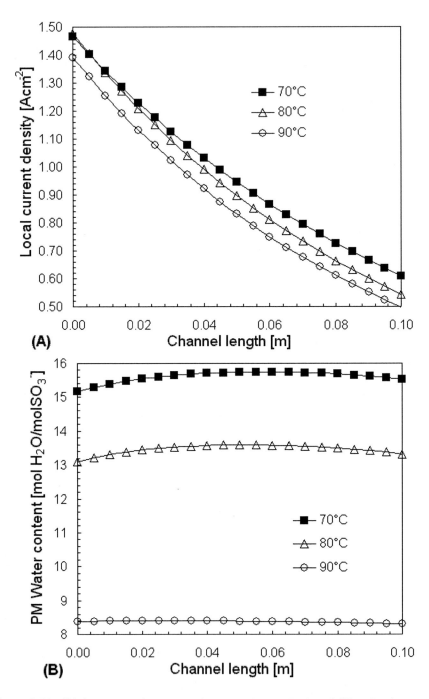

Figure 3.13: (A) Average polymer membrane water content and (B) cathode porous medium saturation along the channel at different operation temperature

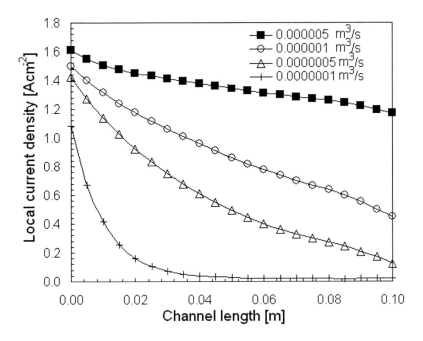

Figure 3.14: Average current density along the gas channel at different cathode inlet volumetric flows

3.3 Numerical study on PEMFC's geometrical parameters under different humidifying conditions

The objective of the present section is to analyze the influences that some geometrical parameters (thickness of polymer membrane, cathode catalyst layer and gas channel to rib width ratio) may have on the water transport and as such on the performance of a straight PEMFC under different hydrating conditions, in order to analyze the diverse phenomena that may dominate electrochemical mechanisms in these systems.

Concentration and ohmic losses may either highly or slightly depend on the geometrical parameters of PEMFC. The geometrical parameters can be of high importance when simulating such systems, consequently their effects should not be taken for granted when either analyzing the water management or validating data. Values of thickness of polymer membrane, cathode catalyst layer and gas channel to rib width ratio were varied between low and high magnitudes under different humidifying conditions (50 and 100%). The effect of gas diffusion layers permeability is also simulated in order to analyze the behavior of concentration losses caused by liquid saturation when using gas diffusion layers with different properties. Results are mostly presented as transversal averages along the gas channel, given that two and three-dimensional displays are not practical to make comparisons. Power density distributions represent the product of current density and cell potential distributions. It is important to mention that the local ohmic losses in the polymer membrane and the overpotential are used to estimate the cell potential distribution. Results are also showed as polarization curves, in order to enhance the presentation of power and current density distributions.

Table D.2 summarizes the base geometrical and electrochemical parameters arbitrarily used in this work. Both cell potential and current density were output variables in simulations. The cathode overpotential is used as controlling input variable, given that both cell potential and current density mainly depend on the concentration and ohmic losses fields in the polymer membrane and cathode catalyst layer. Therefore, their averages and fields are known at end of every simulation. The base value of cathode overpotential at the interface cathode catalyst layer-polymer membrane was kept constant at 0.760 V (when calculating distributions of power and current densities) using the electrochemical parameters arbitrarily chosen in this work. This value represents the reaction resistances in the cathode electrode without considering the concentration and ohmic losses [5] (mainly a property of platinum and external circuit resistances). Therefore, cathode overpotential is assumed constant at 0.760 V for every case (except for the polarization curves), in order to obtain high numerical PEMFC output. Such input data configuration was considered suitable to achieve objectives of the present work. Otherwise, a trial and error procedure would be necessary to achieve results at constant either current density or cell potential which would not be practical when considering both cell potential and current density as output fields. Inlet flows, pressures and temperatures were kept constant. The cathode inlet flow is three times higher than the necessary flow to produce 1.0 Acm^{-2}. Hence, the concentration losses by oxygen depletion are considerably reduced. Each geometrical parameter was varied at constant inlet relative humidity. Afterwards, the same procedure is repeated but using a different inlet relative humidity. Values of inlet relative humidity were 50 and 100%. Even though the stoichiometry may vary for different inlet relative humidities, its change is assumed as negligible in most of the cases given the excess of reactants.

3.3.1 Variation of polymer membrane thickness

Polymer membrane thickness is an important parameter to study, due to its influence on the ohmic losses by proton flux resistance from anode to cathode. The proton conductivity may be highly affected by the water content of polymer membrane, and as such ohmic losses in this region depend not only on the proton flux and length of resistance, but also on the interfacial water transfer through the polymer membrane during operation. As well the thickness of polymer membrane is an important length in the process of hydration, given that it proportionally defines the amount of water to be transferred to increase the water content, consequently its magnitude may have a nonlinear influence on ohmic losses and as such on the performance of PEMFC.

Fig. 3.15 shows averages of the along-channel power density and ohmic losses in the polymer membrane for different PMT using 50 and 100% of RH in both cathode and anode inlet flows. It can be observed in Fig. 3.15(A) the influence of RH of inlet flows on power density. The magnitudes of power density are lower at 50% RH in comparison to the 100% RH values; given that ohmic losses in CCL and PM considerably increase for lower RH as shown in Fig. 3.15(B). The PEMFC power density improves for thinner membranes, given that the ohmic losses in PM are considerably decreased (Fig. 3.15(B)) for both RH cases. The ohmic losses at 50% RH seem to be uniform along the gas channel, on account of the compensation between the gradual humidification of polymer membrane and the increase of current density along the gas channel. This increase of current and power density is produced by the decrease of ohmic losses in CCL and PM as result of the humidification of the polymer electrolyte. Under fully humidifying conditions, the lack of oxygen depletion causes uniform polymer membrane voltage loss along the gas channel. In conclusion, it can be seen in Fig. 3.15 that the ohmic losses can be considerably reduced when using membranes thinner than 50 μm for both 50 and 100% RH.

Fig. 3.16 shows the polarization curves for different polymer membrane thickness at 50 and 100% RH, in order to improve the former observations. The ohmic losses seem to be considerably diminished at 50% when using thin polymer membrane (e.g. 25 and 50 μm). The differences between 100% RH (fully humidifying conditions) and 50% (dehydrating condition) are not dramatical when using such thin membranes. Therefore, suitable operation can be achieved under such conditions. These results show the importance of numerically taking into account every effect that may influence the water transport, in order to determine the actual ohmic and concentration losses and reliably predict the actual PEMFC behaviors.

Fig. 3.17 shows 100 μm thick polymer membrane water content fields (base case) at different inlet flow RH (50 and 100% RH, respectively) to provide insight into water distributions in the polymer membrane. It can be seen in Fig. 3.17(A) that the polymer membrane is considerably dried at the entrance of the gas channel when operating at 50% RH. The ohmic losses reduce the cathode reaction rate and so the water production is not high enough to effectively increase the along-channel water content (approximately from 4 to 7 $(\mathrm{kmol}_{H_2O}(\mathrm{kmol}_{SO_3^-})^{-1})$). At 100% RH (Fig. 3.17(B)), the water content is almost uniform along the gas channel. Even though the water fraction increases along the gas channel, the water content tends to be slightly higher at the entrance as product of the higher cathode reaction rate in this area. This behavior means that most of the water is more likely to be disposed by the gas and liquid phase than to be transferred to the polymer electrolyte under fully humidifying conditions. Therefore the polymer membrane water content may be almost uniform along the gas channel under fully humidifying conditions as a result of the water saturation in the

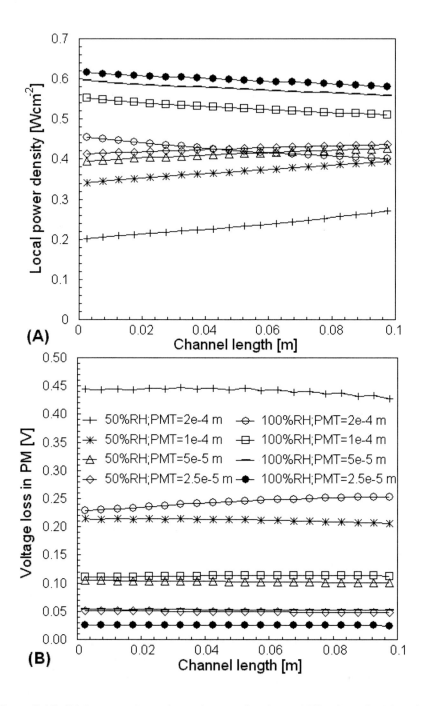

Figure 3.15: (A) Average along-channel power density and (B) voltage loss in polymer membrane for different polymer membrane thickness and inlet flow relative humidities.

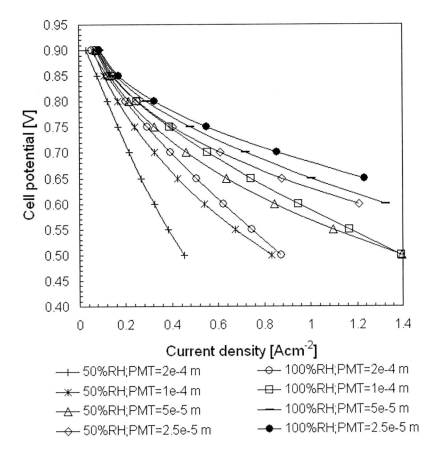

Figure 3.16: Cell voltage vs. average current density for different polymer membrane thickness and inlet flow relative humidities.

control volume (activity 1). The electro-osmotic drag tends to dry the anode side of PM especially at 50% RH. However, fully humidifying conditions in the anode inlet flow (in case B) and the back diffusion contribute to keep water content of PM in anode side.

Back diffusion is an important factor that may be affected by the polymer membrane thickness. Hence, the polymer membrane thickness is a parameter that may nonlinearly affect the performance of PEMFC. Fig. 3.18 shows two contours of water content along the gas channel at 50% RH, in order to provide insights to this factor for PMT (A) 200 μm and (B) 50 μm. Average current density is 0.8 Acm^{-2} for both Fig. 3.18(A) and (B). As it can be seen in Fig. 3.18(A), the water content considerably diminishes from cathode to anode as a product of the electro-osmotic drag and the long path for diffusion of water molecules from cathode to anode side. This effect is considerably decreased for thinner membranes, given that back diffusion becomes

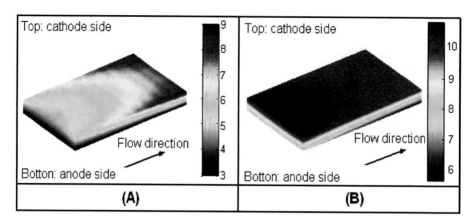

Figure 3.17: Polymer membrane water content field at different inlet flow relative humidities A(50%) and B(100% RH).

important in comparison to the electro-osmotic drag, as a product of the shorter path from cathode to anode side (Fig. 3.18(B)). Fig. 3.18 shows the high negative effect of the polymer membrane thickness over back diffusion and as such over ohmic losses in PEMFC.

3.3.2 Variation of cathode catalyst layer thickness

Catalyst layer thickness defines the length for the proton transport through the polymer electrolyte to active sites in the cathode. Ohmic losses in this region depend directly on this parameter. Dehydrating conditions may affect the cathode reaction rate for diverse CCLT. For these reasons the catalyst layer thickness is varied maintaining the catalyst loading per area (1.0 mgPtcm^{-2}) constant under different humidifying conditions. Even though, the change of catalyst layer thickness causes a variation of catalyst concentration, the total mass of catalyst in the system remains constant.

Fig. 3.19 shows averages of the along-channel current density for different CCLT using 50 and 100% of RH in both cathode and anode inlet flows. As shown in Fig. 3.19, ohmic losses in CCL considerably decrease for thinner CCLT. Thinner catalyst layers enhance cathode reaction rate diminishing resistance to transport of proton to reactive sites. Therefore, thin CCLT favors higher current densities; consequently the performance of PEMFC is increased. This enhancement is more important at dehydrating conditions (50% RH). As it can also be seen in Fig. 3.19, current density is considerably improved for the two thinnest CCL at 50% RH, given that current density distributions are considerably similar at 50 and 100% RH. Hence, ohmic losses in CCL could be considered insignificant for these values. It is important to mention that the slight difference of current density observed for the thinnest CCL is a product of the increase of stoichiometry caused by higher oxygen concentration at 50% RH.

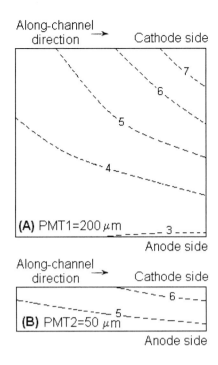

Figure 3.18: Polymer membrane water content at 50% RH A(PMT=200 micrometer) and B(PMT=25 micrometer).

However, this observation do not change the fact that ohmic losses are rather small for thinner CCL than 6.25 μm.

Polarization curves were built to show the whole performance of PEMFC (ohmic losses in CCL and PM) for different CCLT at 50 and 100% RH (Fig. 3.20) and to complete the former observations. As shown in Fig. 3.20, ohmic losses in CCL are considerably reduced when using thinner electrodes (less than 6.25 μm) at different RH which is in agreement with former observations (Fig. 3.19). Given that the polarization curves of the two thinnest CCL almost overlap for both RH, it can be concluded that ohmic losses in CCL may be considered as negligible for these values. Therefore, the difference of performance observed in Fig. 3.20 at different RH is mostly caused by ohmic losses in polymer membrane (base case: 100 μm).

These results show the importance of including CCLT in simulation of PEMFC, even at 100% RH. CCL should be considered as an active volume and not as an active surface, in order to account for ohmic losses caused by polymer contained in this region. Considering CCL as a volume would enhance analyzes concerning PEMFC optimization.

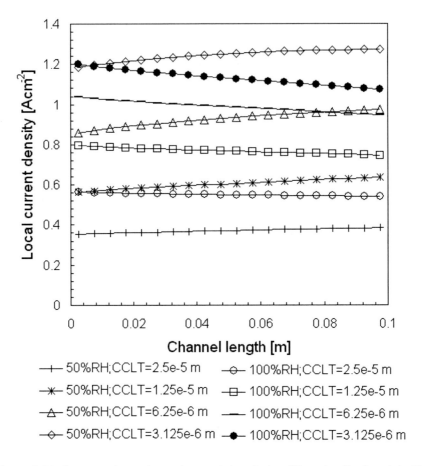

Figure 3.19: Average along-channel current density for different cathode catalyst layer thickness and inlet flow relative humidities.

3.3.3 Variation gas diffusion layer permeability

Gas diffusion layer permeability is an important parameter that may considerably affect PEMFC performance under fully humidifying conditions. In case the gas diffusion layer permeability is small, water saturation may considerably increase and so power density may be affected by concentration losses caused by resistance to reactant transport to active sites. In numerical terms, this factor dominates the magnitude of saturation by controlling the capillary flow to be disposed to the gas channels.

Fig. 3.21(A) and (B) show the effect of gas diffusion layer permeability on power density (polarizations) and water saturation, respectively. The performance of PEMFC considerably decreases when permeability is low (Fig. 3.21(A)), given the considerably increase of concentration losses produced by the liquid saturation (Fig. 3.21(B)). The porosity of gas diffusion layer is kept constant, consequently the behavior ob-

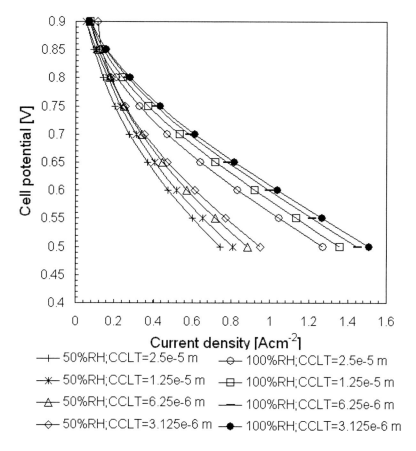

Figure 3.20: Cell voltage vs. average current density for different cathode catalyst layer thickness and inlet flow relative humidities.

served in Fig. 3.21 is only caused by changes of the tortuosity of the path for disposal of liquid water. The power density tends to be constant for permeabilities higher than $2.5.10^{-15}$ m^2 under high current densities, which means that the saturation is considerably controlled over this value (less than 0.2). These results show how the PEMFC performance may be affected by the liquid saturation and the importance of two-phase flow simulation to reduce the concentration losses caused by the water condensation.

3.3.4 Variation of cathode gas channel width

Gas channel width represents a characteristic length perpendicular to former lengths. The relation between gas channel and rib width may be of importance, given that the reactants flow through the gas channel and electrons flow through the ribs. Therefore, both lengths are highly important to be considered when analyzing the concentration

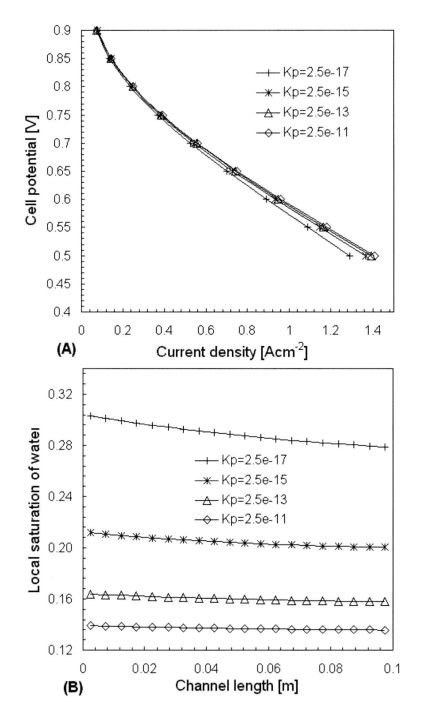

Figure 3.21: A) Cell voltage and average current density and (B) water saturation assuming different cathode permeabilities.

and ohmic losses. However, the resistance to electron transport through ribs is assumed as negligible in this work.

Gas channel to rib width ratio is varied from 5.0 to 1.0, in order to analyze the behavior of the concentration losses under these conditions. As shown in Fig. 3.22, the polarization curves overlap for 50% RH. Hence, the effect of GCW is rather small at 50% RH, provided that the ohmic losses dominate over concentration losses induced by narrow gas channel, and so the oxygen diffusion accomplishes to supply the reactant demand under such low cathode reaction rates. Oxygen diffusion is also more effective, given the higher oxygen concentration at dehydrating conditions. The concentration losses at 100% RH increase in comparison to ohmic losses. The increment of humidifying conditions reduces the negative effect of polymer electrolyte proton resistance and as such the cathode reaction rate is favored. Under these conditions, the intensity of reaction is higher and as such is the demand of reactants. To satisfy this demand, the reactant transport must be effective, but the characteristics length like channel width may exert control over the mass transfer. For these reasons the polarizations show slight differences for diverse GCW at 100% RH where the wider gas channels tend to enhance the performance of PEMFC by increasing the area for diffusion of species. As seen in Fig. 3.22, effect of gas channel width may only be important at high current densities under fully humidifying conditions, given the importance of diffusional control under these conditions. Otherwise, the effect of this parameter can be ignored.

Liquid saturation may not play an important role in these former observations. Fig. 3.23 shows along-channel distributions of water saturation for different GCW at 100% RH. As it can be seen in Fig. 3.23, wider channels tend to diminish the saturation in porous mediums, which is a result of major elimination of liquid water by capillary flow. However, the change of water saturation for different GCW may be considered as negligible when analyzing the effect of the gas channel width on PEMFC performance. Under conditions favoring liquid accumulation in gas diffusion layer, gas channel width may play a more important role. As seen in former section, liquid saturation is likely to be highly governed by parameters like the permeability of gas diffusion layer. Liquid saturation is also dominated by the hydrophobicity of the gas diffusion medium, which is assumed optimal in this work (contact angle$\simeq 110$).

3.3.5 Effect of ohmic losses on temperature and saturation distributions

Temperature distributions are mostly affected by distributions of ohmic losses in PEMFC. If ohmic losses were low and almost uniform along the gas channel, the system could be assumed as isothermal. As seen before, ohmic losses play an important role for thick polymer membranes, which effect is considerably decreased when using thinner membranes. Given that the optimal configuration of geometrical parameters considerably reduces the overall ohmic losses even at high current densities and dehydrating

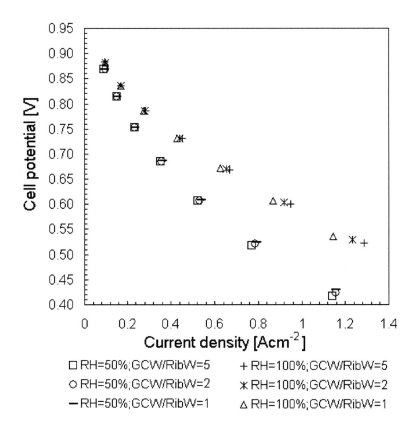

Figure 3.22: Polarization for different gas channel width and inlet flow relative humidities.

conditions, isothermal operation may be assumed in most cases. Temperature distribution may play an important role in a system with serious ohmic losses (e.g. thick membranes greater than 50 μm). Fig. 3.24 shows temperature and saturation contours for two different polymer membrane thickness ((A),(B): 200 and (C),(D): 50 μm, respectively) when operating at 100% RH, in order to determine the importance of temperature distributions caused by ohmic losses under fully humidifying conditions. The heat is mostly produced by the overpotential at cathode catalyst layer and the Joule heating in the polymer membrane. Therefore, the temperature in gas diffusion layer is likely to show the same tendency as the local current density. In this case temperature distributions are almost uniform as a result of the high convection at gas channel-gas diffusion layer interface and low ohmic losses under fully humidifying conditions (Fig. 3.24). However the effect of Joule heating in polymer membrane on temperature and saturation distribution can be observed in Fig. 3.24. Temperature increases for thicker membranes as a product of higher Joule heating and dimin-

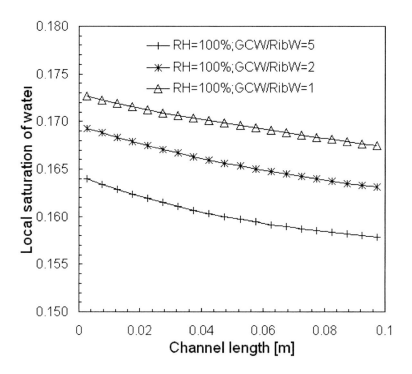

Figure 3.23: Average along-channel water saturation for different gas channel width.

ishes along the gas channel due to the gradual reduction of cathode reaction rate
(Fig. 3.24(A)). In direction from polymer membrane to gas channel, the temperature is
almost uniform due to the small path to dispose the excess of heat to the gas channel
flow. Due to the small residence time of the gas flow, the gas channel temperature may
considered as uniform, and so the temperature in the along-channel direction shows
the decreasing tendency observed in Fig. 3.24(A). Water saturation basically depends
on water activity and temperature. At the gas channel entrance where temperature is
higher, water saturation shows an increasing tendency related to the water production
in cathode and decrease of temperature along the gas channel (Fig. 3.24(B)). However, the water saturation reaches a maximum and starts diminishing as a result of
the lack of water production. As shown in Fig. 3.24(C) and (D), the temperature is almost uniform along the gas channel (Fig. 3.24(C)) and only a slight decrease of water
saturation along the gas channel is observed (Fig. 3.24(D)). As seen in Fig. 3.24(B)
and (D), cathode water saturation is highly dominated by capillary flow and values do
not overcome the magnitude of 0.2, given the high permeability of base case (1.10^{-13}
m^2). Behavior of temperature distributions at 50% RH is assumed to be similar to the
case of 100% RH, except for the case of thick membranes where ohmic losses are
of considerable importance, especially at dehydrating conditions. Given that ohmic

losses are considerably reduced when using thin membranes and cathode electrodes (Fig. 3.15 and 3.16), temperature distributions at different hydrating conditions (50 and 100% RH) are assumed to be similar for systems with low ohmic losses.

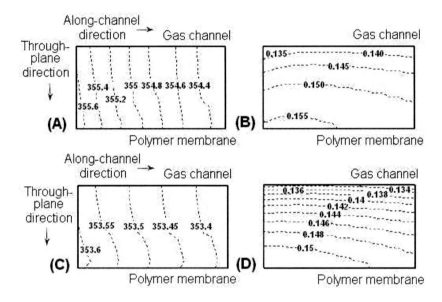

Figure 3.24: Through-plane and along-channel distributions of temperature and liquid saturation (A),(B) PMT=200 micrometer and (C),(D) PMT=50 micrometer under fully humidifying conditions (100% RH).

3.3.6 Polarization at different operational temperatures

Fig. 3.25 shows the polarization curves for different operational temperatures at different humidifying conditions. Temperature seems to play an important role especially at dehydrating conditions. Even though cathode reaction rate is stimulated by higher temperatures, polymer electrolyte water content decreases and so higher resistance to proton flux slows down enhancement of power output as shown in Fig. 3.25 for fully humidifying conditions. This effect becomes dramatical at dehydrating conditions, given the considerable increase of ohmic losses. At low and moderate current density, the tendency is similar for the case of fully humidifying conditions, but performance considerably diminishes for higher temperatures at dehydrating conditions. This observation is important to establish that PEMFC could be significantly affected by temperatures higher than 80 °C when the water and heat management is incorrect. Nonetheless, assuming an scenario of small ohmic losses in the polymer membrane even at dehydrating conditions as shown in Fig. 3.16, effect of temperature may be considerably diminished under dehydrating conditions, and so better performances could be achieved at higher temperatures.

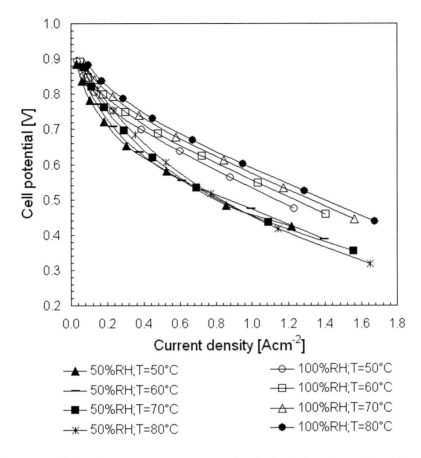

Figure 3.25: Cell voltage vs. average current density for different operational temperatures at different inlet flow relative humidities (50 and 100% RH).

3.3.7 Partial conclusions II

Results of this section showed that geometrical parameters may have an important influence on PEMFC performance under conditions arbitrarily chosen in this work. This work also showed the importance of simultaneously calculating water absorption and desorption by polymer electrolyte and water saturation, in order to establish actual ohmic and concentration losses under different hydrating conditions.

Polymer membrane may cause serious ohmic losses if thickness is high. Use of thin polymer membrane considerably enhances PEMFC power by reducing ohmic losses, especially under dehydrating conditions.

PEMFC power may be improved by using thin cathode catalyst layers. Proton resistance to active sites is small; consequently cathode reaction rate is not affected by activation losses. Not only ohmic losses may be enhanced but also concentration losses. Concentration of platinum particles becomes higher, and so reactants transfer control may be reduced. Therefore cathode catalyst layer thickness plays an important role when optimizing electrochemical parameters.

Permeability of the gas diffusion layer is an important factor to control the liquid saturation. In numerical terms, high permeabilities (greater than 1.10^{-15} m^2) could enhance disposal of liquid water in the gas diffusion layer, keeping the saturation in porous mediums controlled (less than 0.2).

Wide gas channels yield lower concentration losses under fully humidifying conditions, and as such better performance is achieved under these conditions at high current densities. However, ribs width should be wide enough to avoid ohmic losses by electron transport from anode to cathode.

Temperature gradients could be important when current density causes high ohmic losses. Otherwise, the system could be treated as isothermal under the electrochemical and geometrical conditions of this work.

Effect of the humidity over PEMFC power density could be considerably diminished if using the optimal geometrical parameters to minimize ohmic and concentration losses. Operation at 50% RH showed to be satisfactory when using polymer membranes thinner than 50 μm. In the case of catalyst layers, PEMFC performance is likely to become independent of this factor if using a thickness thinner than 6.25 μm even at dehydrating conditions like 50% RH.

3.4 Concentration and ohmic losses in free-breathing PEMFC

PEMFCs are usually operated under humidifying conditions around 80 °C to avoid dehydration of polymer electrolyte, reduce water condensation and increase reaction rate. Hence compressors, humidifiers and heaters are needed to achieve high and optimal power densities. Nevertheless, PEMFC may also be used for portable applications in which additional accessories like humidifiers and compressors can not be utilized for practical reasons. Thus optimal performance of these applications directly depends on how species are supplied and disposed in the MEA (membrane and electrode assembly). Such applications are called "free-breathing" PEMFC provided that oxygen is taken from ambient air by diffusion and natural convection.

Performance of free-breathing PEMFC depends highly on natural convection and oxygen diffusion and so concentration losses become high in comparison to ohmic losses. The slow motion of reactants and oxygen diffusion are not the only factors that may cause high limiting effects, low convection may also result in a build up of water causing severe flooding at the cathode.

Even though concentration losses in cathode dominate free-breathing PEMFC performance, calculation of water content of polymer electrolyte and liquid saturation is necessary to establish a complete ohmic-concentration losses analysis. Previous experimental works concerning free-breathing PEMFC have been presented in [13, 51–61]. Mennola et al. [57] presented both experimental and numerical data of oxygen and water mole fraction fields as well as velocity fields to provide insight of many details that may be important in understanding mass transport limiting controls in free-breathing PEMFC.

Given the small characteristic lengths in PEMFC, numerical modeling is important to provide insight into the different phenomenon and predict distributions that are difficult to be experimentally determined. Hence the objective of the present section is to carry out simulations to quantitatively analyze ohmic and concentration losses of free-breathing PEMFC power by using a non-isothermal and three-dimensional model to account not only for natural convection and oxygen diffusion but also for water and heat management. Flow fields, energy and species transport, transport of water in polymer membrane and movement of liquid water in cathode and anode porous layers are determined under different hydrating conditions, cell temperatures, cathode catalyst loadings and channel lengths, in order to numerically determine limiting controls and propose solutions for enhancement of free-breathing PEMFC performance.

Natural convection is introduced in cathode gas channel momentum equation as source term (Table 2.1)

$$ST_V = \varrho \beta g(T - T_{amb}) \tag{3.1}$$

Inlet anode flow temperature is assumed equal to cell temperature, and so numerical results of temperature become almost uniform, even though cathode inlet flow temperature is constant in all calculations (25 °C).

As mentioned above, catalyst layers are assumed as an active volume, and so cathode overpotential becomes an output variable due to proton resistance. Hence, input cathode overpotential at the cathode catalyst layer-polymer membrane interface must be defined, in order to estimate ohmic losses in the cathode catalyst layer. This reference cathode overpotential is assumed constant (0.8 V under electrochemical conditions arbitrarily chosen in this work). Catalyst loading is assumed as 1.0 $mgcm^{-2}$ and catalyst surface area as 250000 (cm^2g^{-1}). Both values may provide relative good performance under free-breathing conditions, and so current density distributions may be suitable to study cell output under diverse conditions. It is important to mention that both values do not represent optimal electrode characteristics, because they were arbitrarily chosen. To analyze optimal electrode characteristics the relationship between catalyst loading and surface area should be defined, in order to carry out numerical studies concerning catalyst loading in free-breathing PEMFC. Such task is not an objective of present work.

Dimensions of control volume were chosen slightly different than former cases, given the limiting effects of free-breathing applications. Table 3.2 summarizes geometrical, electrochemical and operating parameters arbitrarily used in this section about free-breathing PEMFC.

Table 3.2: Geometrical and electrochemical parameters of free-breathing PEMFC

Description	Value
channel height	0.003 (m)
channel width	0.003 (m)
channel length	0.05 (m)
channel rib	0.0002 (m)
gas diffusion layer thickness	0.0001 (m)
gas diffusion layer porosity	0.5
catalyst layer thickness	10 μm
catalyst layer porosity	0.4
polymer membrane thickness	230 μm
Pt loading	1.0 $(mg_{Pt}cm^{-2})$
Pt surface area	250 $(cm^2_{Pt}mg_{Pt}{}^{-1})$
agglomerate porosity	0.2
agglomerate radii	0.00001 (cm)
Inlet Anode Volumetric Flow	0.0000001 (m^3/s)
Inlet Cathode Flow Pressure	1 (atm)
Inlet Anode Flow Pressure	3 (atm)
Ambient temperature	25 (°C)

3.4.1 Effect of humidity conditions

As oxygen is directly taken from surrounding air, ambient conditions may have an influence on free-breathing PEMFC performance. Ambient relative humidity is taken as 40, 70 and 100% at 25 °C for different cell temperatures (25, 50 and 75 °C) while anode inlet flow is assumed to be fully humidified (100% ARH) at cell temperature.

Adjusting cell temperature at 25 °C (gas channel walls temperature), average power density is 0.0245, 0.0251 and 0.0244 Wcm^{-2} (approx. at 0.58 V) for 40, 70 and 100% air relative humidity, respectively. In this case oxygen supply is rather small, given that natural convection is negligible and diffusion through ends of cathode gas channel is the unique supply of oxidant for electrochemical reaction. Current density is rather low as a result of slow diffusion, and so cell potential is almost constant for every ambient relative humidity. As shown in Fig. 3.26, concentration losses highly dominate performance under these conditions, given that current density curves overlap for different ambient relative humidities. Therefore ambient relative humidity exerts a negligible effect on ohmic losses as result of low current densities yielded under such oxygen transfer control, and so power output is only governed by oxygen diffusion. Most of oxygen is consumed near ends of cathode gas channel and so average power density becomes low and highly controlled by oxidant transport. A considerable portion of the active area is barely used due to poor axial oxidant transport against through-plane diffusion and activation potential. Highest current densities at ends of cathode gas channel having almost symmetrical magnitudes are product of negligible convection and oxygen diffusion through boundaries.

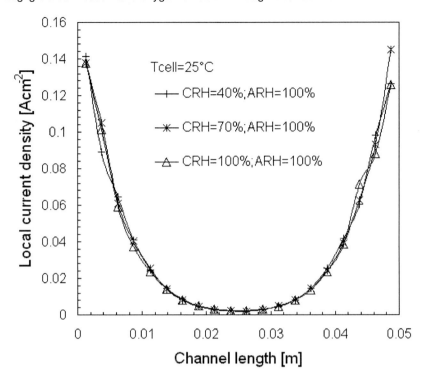

Figure 3.26: Along channel current density for different ambient relative humidities at 25 °C.

At 50 °C of cell temperature, average power density is 0.0451, 0.0454 and 0.0460 Wcm^{-2} (approx. at 0.56 V) for 40, 70 and 100% air relative humidity, respectively. Considering that cell temperature is higher than ambient temperature, natural convection yields a smooth stream through cathode gas channel improving oxygen supply and as such the performance of the cell. The average inlet air velocity through gas channel is 0.017 ms^{-1}. As it can be observed in Fig. 3.27, highest current density is seen at entrance of gas channel. Oxygen is rapidly consumed by electrochemical reaction given the high activation potential needed to operate around 0.6 V. Presence of natural convection improves performance due to major availability and transport of oxidant in comparison to only-diffusion-driven-flux and then current density increases in comparison to former case (Fig. 3.26). Even though natural convection enhances oxygen supply, the cathode reaction seems to be highly controlled by concentration losses, given that cell potential is almost constant for every air relative humidity, which means that ohmic losses may be negligible under these conditions. At upper end of gas channel, oxygen counter-diffusion (against natural convection) yields a slight increment of current density as product of the short path between gas channel end boundary and active sites of cathode catalyst layer.

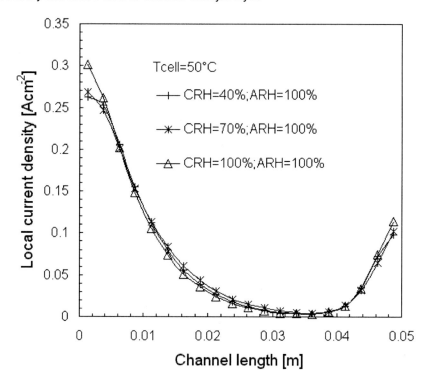

Figure 3.27: Along channel current density for different ambient relative humidities at 50 °C.

Increasing the cell temperature to 75 °C, average power density is 0.0633, 0.0629 and 0.0625 Wcm^{-2} (approx. at 0.56 V) for 40, 70 and 100% air relative humidity, respectively. Fig. 3.28(A) shows overlapping curves for different ambient relative humidities, which means that performance is even highly controlled by mass transport deficiencies over possible dehydration of polymer electrolyte. Nevertheless, current densities increase in comparison to former cases (25 and 50 °C), basically due to increment of natural convective flow (0.030 ms^{-1} average inlet air velocity) and oxygen diffusion. Not only natural convection and oxygen diffusion play an important role in electrochemical reaction but temperature does as well, given the increase of activation of cathode reaction by temperature. In this case, polymer membrane voltage loss is showed for different ambient relative humidities (Fig. 3.28(B)) in order to observe role of ohmic losses on cell performance under these conditions. It can be seen from Fig. 3.28(B) that voltage loss curves overlap for different ambient relative humidities, which means that ambient relative humidity may not affect cell output.

Nonetheless, anode inlet flow relative humidity may have an important influence

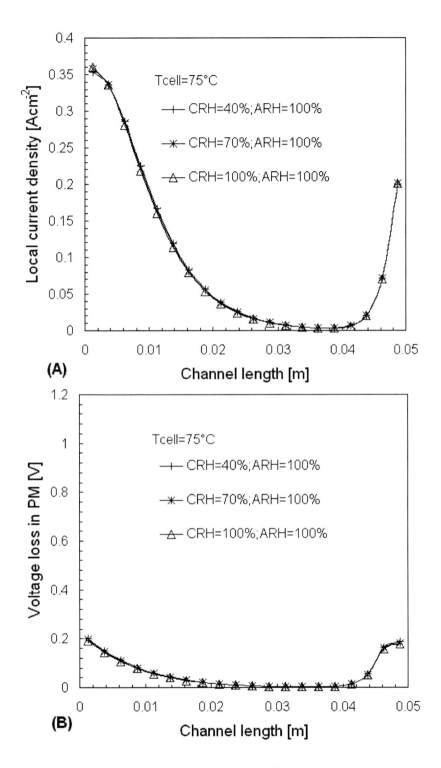

Figure 3.28: A) Along channel current density and (B) voltage loss in polymer membrane for different ambient relative humidities at 75 °C.

on hydration of polymer membrane provided that water content curves almost overlap for different ambient relative humidites (Fig. 3.29). Therefore fully humidifying conditions of anode inlet flow may reduce effect of ambient humidifying conditions on cell output by effectively hydrating polymer electrolyte. Hence the effect of anode inlet flow relative humidity becomes of importance when analyzing possible ohmic losses in free-breathing PEMFC.

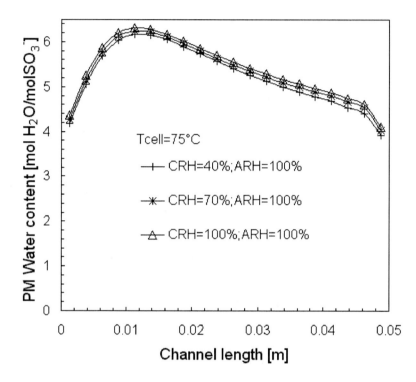

Figure 3.29: Along channel average polymer membrane water content for different ambient relative humidities at 75 °C.

For this reason, anode inlet flow relative humidity was varied from 10 to 100% keeping ambient relative humidity at 70%, in order to observe performance of free-breathing PEMFC under such dehydrating conditions. As it can be seen in Fig. 3.30(A), anode inlet flow relative humidity may considerably affect cathode oxidation reaction by dehydrating the whole polymer electrolyte, and so increasing ohmic losses in cathode catalyst layer. Even though current density is relatively low, dehydration of polymer electrolyte may increase ohmic losses, consequently current density is affected at entrance of gas channel by different anode inlet flow relative humidities (Fig. 3.30(A)). These differences decrease along gas channel due to depletion of oxygen and so irreversible local losses become minimal as product of low local current densities. This behavior can also be observed in polymer membrane voltage loss curves (Fig. 3.30(B)). Even though performance of free-breathing PEMFC may be slightly affected by such dehydrating conditions, concentration losses are definitely even more important. Average power density is 0.041, 0.040, 0.043 and 0.045 Wcm^{-2} (approx. at 0.54 V) for 10, 40, 70 and 100% anode inlet relative humidity, respectively. Therefore, ohmic losses become slightly important when analyzing free-breathing PEMFC as long as local current densities are over 0.15 Acm^{-2} as it can be observed in Fig. 3.30(A) and (B).

3.4.2 Mass transport control

As mentioned above, oxygen supply is highly limiting in free-breathing PEMFC. Either with or without natural convection, oxygen transport to active sites seems to be rather poor as seen in former results. For this reason, estimation of oxygen concentration field is of high importance to illustrate limiting conditions to cathode electrochemical reaction. Fig. 3.31(A),(B) and (C) show along-channel and through-plane oxygen mass fraction maps using fully humidifying conditions at 25, 50 and 75 °C, respectively. The data aspect ratio was changed by reducing channel length aspect (1:5). Results show that through-plane oxygen transport is more effective in comparison to axial transport for all cases (25, 50 and 75 °C). Most of oxygen is consumed near gas channel ends given the more effective through-plane diffusion in comparison to axial convective and diffusive oxygen transport. Such behavior may be caused by differences in characteristic lengths for cathode mass transfer. Through-plane and axial characteristic lengths are around 0.003 and 0.05 m, respectively. Therefore short path of through-plane oxygen diffusion and activation of cathode reaction deplete oxygen rapidly and so a considerable portion of active sites may be barely used under these conditions given the relative long characteristic length for axial oxygen transport. Lowest cell temperature (25 °C) and negligible natural convection considerably limit cell performance to low current densities as it can be seen in Fig. 3.31(A). Oxygen mass fraction is rapidly diminished at ends of gas channel from 0.23 to 0.12, given the high oxygen demand in this region. Afterwards axial diffusion of oxygen is rather slow; consequently oxygen mass fraction highly diminishes towards center of control volume. Fig. 3.31(B) show that natural convection may enhance oxygen transport to active sites. Oxygen mass fraction is higher along gas channel entrance in comparison to Fig. 3.31(A), basically due to increase of natural convection. Consequently, oxygen depletion increases as a product of higher cathode reaction rate, which is stimulated by higher temperature and oxygen concentration. Therefore, the mass fraction map at 25 °C (Fig. 3.31(A)) is similar to the 50 °C map at the center of the cell (Fig. 3.31(B)). Towards the end of the gas channel, natural convection dominates transport of oxygen over oxygen diffusion and so Fig. 3.31(B) shows a considerable decrease of oxygen fraction at upper end of gas channel in comparison to Fig. 3.31(A), given that upstream depleted oxygen flow is much higher than oxygen

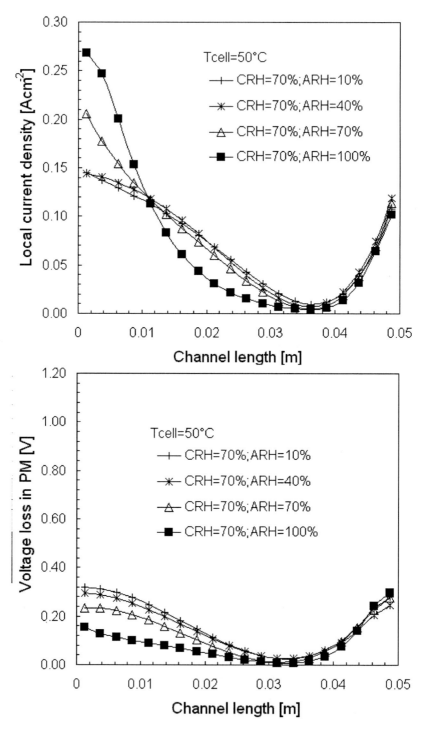

Figure 3.30: A) Along channel current density and (B) voltage loss in polymer membrane for different anode inlet flow relative humidities at 50 °C.

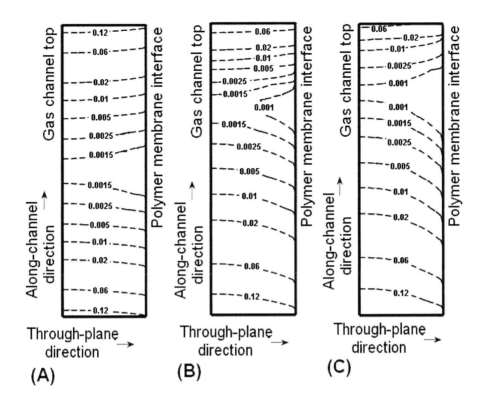

Figure 3.31: Through-plane and along-channel oxygen mass fraction distributions at (A) 25, (B) 50 and (C) 75 °C.

counter-diffusion, and so lower oxygen mass fraction is observed. Even so, performance of cell increases at higher temperatures. Fig. 3.31(C) showed to be similar to Fig. 3.31(B), even though temperature is higher, and so oxygen consumption. This effect is explained by former observation. Even though natural convection and diffusion are enhanced at 75 °C in comparison to 50 °C, cathode reaction rate is also higher and so oxygen depletion. As mentioned before, a lower oxygen fraction at upper end of gas channel (Fig. 3.31(B) and (C)) is a product of higher oxygen depleted upstream in comparison to counter-diffusion.

As seen cell temperature affects performance by increasing natural convection, oxygen diffusion and activation potential. Fig. 3.32(A) and (B) illustrate the role of temperature on cell output under fully humidifying conditions. Cell performance can considerably be increased using higher temperatures (Fig. 3.32(A)). Ohmic losses show a negligible effect on cell output (Fig. 3.32(B)), which means that concentration losses govern cell performance under free-breathing conditions.

Under such mass transport limitations, cathode catalyst loading may be an important parameter to observe. Fig. 3.33 shows current densities distribution for different

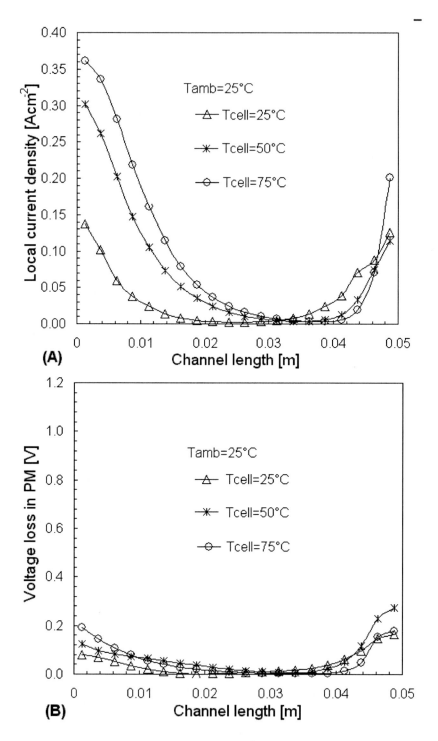

Figure 3.32: Along channel current density and (B) voltage loss in polymer membrane for different cell temperatures.

cathode catalyst loadings (1.0, 2.5, 5.0 and 10 mgcm^{-2}). As it can be seen in Fig. 3.33, cathode catalyst loading enhances current density at gas channel ends. However this effect diminishes for higher catalyst loading, due to poor axial diffusion. Differences in current density distributions dramatically decrease along the gas channel, as result of depletion of oxygen which exclusively dominates performance over increase of active sites. Therefore increasing catalyst loading under such mass transfer controls may represent a high loss of cathode reaction efficiency.

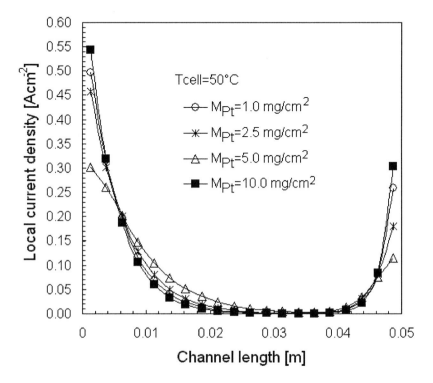

Figure 3.33: Along channel current density using different cathode catalyst loadings at 50 °C.

Considering such limitations of axial oxygen transport, cell axial length is reduced to observe improvements in oxygen transport and overall performance. Results of Fig. 3.31, 3.32 and 3.33 showed that cathode feed may be rather slow, and so free-breathing PEMFC performance is rather limited by rapid oxygen depletion at electrical potential around 0.6 V. Fig. 3.34 shows current density distributions using different channel lengths. As it can be seen, magnitudes of average current density are highly different due to shorter axial diffusion path, and so the active area becomes better utilized under chosen conditions. At end of gas channel, current density varies as product of upstream oxygen flow which is not depleted for shorter gas channels. End current density is smallest for longest channel, given the depletion of oxygen upstream. However, once channel length is shorter, oxygen upstream and counter-diffusion yield higher current densities towards end of channel. A simple approach

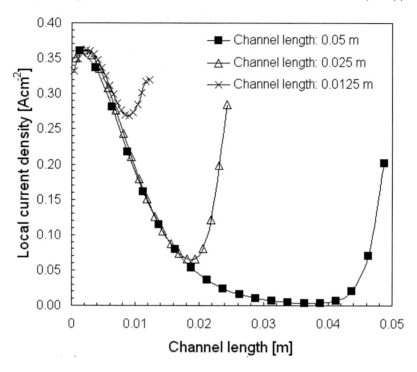

Figure 3.34: Along channel current density using different channel lengths at 75 °C.

can be made, in order to demonstrate role of natural convection over oxygen diffusion on concentration losses. Cathode reaction rate may be assumed as a first order mechanism, in order to simplify the analytic expression:

$$r_{O_2} = K_1 C_{O_2}^{CL} e^{-\frac{K_2}{T}} \tag{3.2}$$

where K_1 and K_2 represent cathode kinetic constants. Concentration of oxygen in cathode catalyst layer can be related to oxygen concentration at inlet as follows:

$$C_{O_2}^{CL} = \frac{C_{O_2 inlet}}{\left[1 + K_1 e^{-\frac{K_2}{T}} \left(\frac{\delta_{GDL}}{D_{O_2,m}} + \frac{L_{GC}}{H_{GC} V_{inlet}} \right) \right]} \tag{3.3}$$

Terms that represent oxygen transport to active sites in former expression (Eq. 3.3) are:

$$\left(\frac{\delta_{GDL}}{D_{O_2,m}} + \frac{L_{GC}}{H_{GC} V_{inlet}} \right) \tag{3.4}$$

Convective term (right) is around 1000 times higher than diffusive term (left) at 50 and 75 °C, which means that oxygen concentration in cathode catalyst layer is basically governed by behavior of natural convection under conditions of this work. Therefore, role of temperature on oxygen diffusion may not be of importance when analyzing concentration losses, given that axial oxygen transport dominates free-breathing PEMFC performance.

3.4.3   Partial conclusions III

In this work results showed that dehydrating conditions may yield considerable ohmic losses only if oxygen transport is high enough to produce average current densities over 0.15 Acm^{-2}. Therefore use of fully humidified anode inlet flows would improve free-breathing PEMFC performance in case of operating at moderate current densities (> 0.15 Acm^{-2}). Nevertheless, ohmic losses definitely play a secondary role in performance of free-breathing PEMFC.

Concentration losses dominate free-breathing PEMFC performance. Axial oxygen transport constitutes main power loss, given that oxidant inlet flux is driven by natural convection and diffusion, which may be considered as slow transport mechanisms in comparison to forced convection. Therefore oxygen availability in active sites is highly limited and as such so current density.

Under conditions arbitrarily used to obtain numerical results, most of active area was barely utilized, given the unbalanced relation between axial transport of oxygen and activation of cathode reaction. Then considering such oxygen transport control, channel length constitutes an important parameter to account for, because most of reaction may take place at channel ends.

Considering high oxygen transport resistance in comparison to activation potential, only small amounts of cathode catalyst loadings are needed to yield acceptable free-breathing PEMFC performances.

Wider gas channels may also enhance performance of free-breathing PEMFC by improving axial transport of oxygen. However, mechanical support of fuel cell stack is also a factor to account for when using wider gas channels.

Chapter 4

SUMMARY

A fuel cell is a device that can convert chemical energy into electricity without me-chanical moving parts. Among the different types of fuel cells, the proton exchange membrane fuel cell (PEMFC) turns up as the most attractive candidate to substi-tute traditional energy converters, given its low temperature of operation $\approx$ 80 °C, higher efficiency and power density. Its range of application is wide and comprises power supply to small electronic devices, transportation and power generation plants. PEMFC performance highly depends on its operational conditions, and so water and heat management are needed to achieve high power outputs. Inlet flows are usually injected as a mixture with water vapor at a certain temperature to avoid the polymer electrolyte dehydration and so alleviate the resistance to proton flux. However, the water produced by the cathode reaction and the lack of diffusion for the disposal of this water may cause condensation in the cathode porous mediums that may clog pores and make difficult the transport of reactants to active sites. Temperature not only stimulates the cathode reaction and increases the saturation pressure but dehy-drates the polymer electrolyte as well. Therefore the temperature plays an important role in water management, consequently water and heat management are closely related. Water and heat management should be properly established to avoid both dehydration of the polymer electrolyte and flooding in the cathode porous mediums. Concentration and ohmic losses must be diminished as much as possible, in order to achieve higher performances at high current densities under a wide range of condi-tions. It is necessary to identify and understand several transport and electrochemical processes that affect the PEMFC power output, in order to enhance and optimize the operation of these devices under different conditions.

Optimization and understanding of PEMFC become a difficult task, given the amount of parameters involved in the whole electrochemical and transport processes, which are differently affected by every variable. Modeling and simulation is the most usefully, fast, inexpensive and practical tool for study and optimization of PEMFC. In the case of PEMFC, numerical simulation can lead to results difficult to obtain by experiments. Nevertheless, modeling and numerical solutions must be focused on representing the dominant processes of PEMFC, in order to obtain more realistic results. Additionally, modeling and simulation of PEMFC should be properly established, in order to obtain insightful analyses.

In order to achieve a complete solution of the transport and electrochemical phenomena in PEMFC, the water management, which in turn requires the consideration of the heat management, must be appropriately determined in terms of modeling and simulation. Hence, the influence of water on PEMFC performance should be properly simulated, and as such the water movement in polymer membrane as well as the condensation and capillary flow in the cathode gas diffusion layer and catalyst layer should be properly estimated. Finally, the flow field in the gas channel should be able to dispose the condensed water. If these aspects are neglected, the numerical results may be erroneous and misleading, given that the behavior of the whole system is highly dependant on liquid and gaseous water transport. Therefore, modeling and numerical simulation of PEMFC should be aimed for simultaneously calculating the water transfer through the polymer membrane and catalyst layers interface as well as the liquid water saturation in porous mediums. PEMFC's simulations under different hydrating conditions become essential when numerically analyzing the water and heat management. Estimations of concentration and ohmic losses of power under different conditions are important for the optimization of PEMFC. Realistic numerical results over a wide range of conditions should be obtained, in order to numerically determine the most limiting factors in PEMFC operation and how they could be diminished. Understanding and predicting of power losses in the PEMFC control volume are imperative to propose solutions and strategies.

Many numerical models have been created to account for water and heat management [4, 6, 7, 25, 27, 29, 30, 34, 35, 37–40]. The modeling aspects considered as highly important by present work can be classified as follows:

- As aforementioned, the diffusion resistance in the non-equilibrium process of water transfer through the polymer membrane interface should be considered. The boundary condition at the polymer membrane interface should be modeled by diffusion in both mediums when using differential modeling. Otherwise the water transfer in and out of the polymer membrane can not be established, and so the water balance in the whole cell may be erroneous. All of the works cited before do not account for any actual water balance in the whole PEMFC system. The interfacial water content of the polymer membrane is calculated assuming equilibrium with the bulk gas phase, which does not consider the species and whole mass balance in the system. Such condition can not establish an actual water exchange between the polymer membrane and the surroundings.

- The equilibrium polymer electrolyte water content at temperatures higher than 30 °C should be estimated, given the dehydration of polymer electrolyte at higher temperatures. The data of Hinatsu et al. [16] at 80 °C can be useful to achieve

78

an estimation of the equilibrium water content of the polymer electrolyte at temperatures between 30 and 80 °C by linear interpolation. Most of the works cited before does not account of the dehydration of polymer electrolyte at temperatures higher than 30 °C, which may cause misleading results.

• When assuming catalyst layers as a surface, ohmic losses in this region are considered as negligible. It is important to model the resistance to the proton flux exerted by the polymer electrolyte contained in the cathode catalyst layer, especially at dehydrating conditions, given that such process may affect the activation potential of the catalyst, and so the sensitivity of the numerical modeling under different conditions may be different. This aspect is usually ignored and the catalyst layers are treated as active flat surfaces without transport resistances.

• Even though concentration losses in anode side are considered insignificant in the present work, liquid water saturation in the anode porous mediums should be included in simulations, given that this phenomenon may influence water transfer through the polymer membrane and anode porous medium interface. In other words, presence of liquid water in anode porous mediums affects the anode water balance, and so a fraction of water may be either absorbed or desorbed by the polymer membrane, another portion may condense and the rest may be disposed by the gas flow in channel. Consequently overall ohmic losses may be affected by both cathode and anode condensation of water, which means that this phenomenon should be taken into account, even though concentration losses in anode are considered as negligible.

Present work aims for the simultaneous calculation of the water transport in polymer membrane (electro-osmotic drag, back diffusion and water interfacial transfer) and liquid saturation in the cathode and anode porous mediums under different operational conditions. It is important to simulate both the polymer electrolyte water transport and the liquid water saturation in order to determine the actual ohmic and concentration losses distributions in PEMFC. Thus the main objective of this work is to present a numerical solution of the water and heat management in PEMFC. The polymer membrane is considered as a component highly affected by its external conditions, hence the mass transfer between the membrane, the cathode and the anode porous mediums is taken into account, in order to determine the actual water content field, so the actual voltage losses in polymer membrane and the cathode electrolyte overpotential are obtained. The liquid water saturation factor in the porous medium is estimated by relations based on the kinetic theory of gases and by the Leverett function for the capillary movement of liquid in unsaturated hydrophobic porous mediums [4, 29, 30, 34]. An electrochemical and phenomenological model is proposed and based on several important phenomena that must not be taken for granted when studying water and heat management in PEMFC. The mathematical solution was adapted to the model and not the model to the mathematical solution. Flow fields, species transport, transport of water in polymer membrane and movement of liquid water in cathode and anode porous layers were determined, in order to accomplish a complete estimation of ohmic and concentration losses of PEMFC power. Output variables are velocity, temperature, mass fraction, current density, voltage loss, water content of the polymer membrane, saturation and liquid flow fields. Effects and behaviors observed from present work are of high importance and interest.

Influences that some geometrical parameters (thickness of polymer membrane, cathode catalyst layer and gas channel to rib width ratio) may have on the water transport and as such on the performance of a straight PEMFC under different hydrating conditions are analyzed, in order to study the diverse phenomena that may dominate electrochemical mechanisms in these systems and improve the performance and utility of PEMFC modeling and simulation. Free breathing PEMFC was also studied in order to quantitatively determine concentration losses under such limited reactants transport conditions. Important conclusions concerning enhancement of PEMFC performance were achieved.

This work can be defined as a numerical solution for the study and analysis of PEMFC components. Behaviours of PEMFC under diverse conditions are observed. Actual driving and limiting factors are pointed out which can be useful for the improvement of the research and development of this technology. Important conclusions derived from numerical results can be presented as follows:

- In the whole, PEMFC should be able to operate under moderate hydrating conditions, in order to make its range of application wider. For this reason, simulations under dehydrating conditions were a primary objective, in order to prove either the suitability or non-viability of PEMFC operation under these conditions. Hence, actual water transfer in and out of the polymer membrane and estimation of activation losses in cathode catalyst layers are factors that cannot be ignored when simulating under dehydrating conditions, given that PEMFC performance highly depends on water transport for these cases. Results showed a "possible" operation without "pre-hydrating" inlet flows (or using moderate relative humidities of inlet flows e.g. 50% for conditions arbitrarily chosen in this work), if proper geometrical parameters were chosen to diminish as much as possible the concentration and ohmic losses. Value of 50% RH is just an example to show the suitability of PEMFC operation under non-fully-humidifying conditions.

- Under fully humidifying conditions, water condensation and saturation of porous mediums showed to be highly controlled by permeability of porous mediums given that liquid flow is mostly driven by capillary forces. In case the permeability is high, water saturation in porous mediums does not affect the performance of PEMFC. However many experimental efforts should be made to improve two-phase modeling in PEMFC. In the first place, parameters like droplet distributions in porous mediums and the behavior of liquid water in gas channel should be properly established in small channels where interfacial forces may play the most important role. Studying two-phase flow may be a too difficult task, given the different flow patterns that can be formed under different conditions. Therefore, its study should be exclusively experimental; important and dominant factors should be put into equations (mechanistics modeling) to perform better simulations when water condensates. Nevertheless, it can be established that the condensation of water is a phenomenon that should be avoided to enhance the application of PEMFC.

- Results show that there may be severe mass transfer limitations depending either on the design or on the water management of the cell. The performance of PEMFC is seriously affected under dehydrating conditions. However, such performance may be considerably improved by using suitable geometrical parameters. These results show the importance of simultaneously calculating both the water absorption and desorption through the polymer electrolyte and the liquid saturation in the cathode and anode porous mediums to obtain an actual view of ohmic and concentration losses of the PEMFC performance.

- Numerical model and simulations proved to be suitable to study diverse phenomenon that limit PEMFC performance. In the whole, PEMFC should be able to operate under moderate hydrating conditions, in order to achieve higher overall efficiencies and avoid concentration losses produced by water saturation at high current densities. PEMFC should reduce as much as possible its "complex" operation (fuel, water and heat management), in order to become more attractive for mass commercialization.

This work was focused on the important factors that affect the PEMFC operation under different conditions. Its range of applications is wide and so a straight PEMFC was assumed as a suitable control volume to quantitatively determine the most limiting factors of performance. Large-scale simulations would be more useful for specific applications and to study heat transfer in and out of the system. In this work, it has been observed that most of the heat produced by power losses was disposed through the axial gas channel flow downstream. Nevertheless, optimizing certain applications of PEMFC would be an interesting task to perform. Fuel management would be an additional factor to account for, given the complexities inherent to hydrogen production. Use of reforming and shifting to produce hydrogen from light hydrocarbons or storage of pure hydrogen in carbon fibers would be an interesting future objective to perform simulations of whole PEMFC systems. With other words, focusing on a certain application (e.g. stationary, transport or small portable devices) is imperative to extend the reach of this work.

The application of PEMFC should be simplified as much as possible to introduce this technology to mass commercialization. Optimization involves not only a set of numerical simulations, but also experimental improvement of polymer electrolyte and electrode performance to achieve better power output using less active areas under simpler operational conditions. In the whole, results of present work were useful to enhance formulation of PEMFC modeling and to evaluate an hypothetical PEMFC operation under moderate hydrating conditions like 50% RH using Nafion as default polymer electrolyte. Results under free-breathing conditions were a complementation to analyze performance of this application which is considerably dominated by axial oxygen transport.

Appendix A

Nernst equation in PEMFC

The change in pure chemical free energy for a half-cell reaction occurring at the standard state can be written as follows [5]:

$$\Delta G^{\text{half}-\text{cell}} = \sum_{i \in \text{React}} \nu_i \mu_i^\circ - n\mu^\circ[e^-(\text{Metal})] \qquad (\text{A.1})$$

The former equation can also be expressed as follows:

$$E_{\text{half}-\text{cell}} = -\frac{1}{nF} \sum_{i \in \text{React}} \nu_i \mu_i^\circ + \frac{\mu^\circ[e^-(\text{Metal})]}{F} \qquad (\text{A.2})$$

where the set of species $i \in$ React contains all the other species taking part in the half-cell reaction, but not electrons which are taken by the second right term in the eq. A.1. Since the species $i \in$ React does not include electrons, they are species whose electrochemical standard potential (expressed in this section and in the literature as μ_i° for each specie) are well defined and tabulated. The chemical potential of an electron $\mu^\circ[e^-(\text{Metal})]$ in a metallic phase is calculated from[5]:

$$\mu^\circ[e^-(\text{Metal})] = -F(\phi_{\text{Metal}} - \chi_{\text{Metal}}) + F(4.44V - \chi_{\text{aq}}) \qquad (\text{A.3})$$

where the term $F(4.44V - \chi_{\text{aq}})$ corresponds to the chemical potential of an electron in standard state $\mu^\circ[e^-(G)]$ and is obtained from a standard hydrogen electrode as follows:

$$
\begin{aligned}
\tfrac{1}{2}\mu^\circ[H_2(G)] - \mu[H^+(\text{aq})] &= (\tfrac{1}{2}\mu^\circ[H_2(G)] - \mu^\circ[H(G)]) \\
&+ (\mu^\circ[H(G)] - (\mu^\circ[H^+(G)] + \mu^\circ[e^-(G)])) \\
&+ (\mu^\circ[H^+(G)] - \mu^\circ[H^+(\text{aq})]) + \mu^\circ[e^-(G)]
\end{aligned}
\qquad (\text{A.4})
$$

the following numerical values are given in the literature:

$(\tfrac{1}{2}\mu^\circ[H_2(G)] - \mu^\circ[H(G)]) = -203.3 \text{kJmol}^{-1}$ [62]
$(\mu^\circ[H(G)] - (\mu^\circ[H^+(G)] + \mu^\circ[e^-(G)])) = -1313.8 \text{kJmol}^{-1}$ [62]
$(\mu^\circ[H^+(G)] - \mu^\circ[H^+(\text{aq})]) = 1088 \text{kJmol}^{-1} + F\chi_{\text{aq}}$ [63]

Additionally $\mu°[H_2(G)]$ and $\mu°[H^+(aq)]$ are set as zero point at the standard state. The expression used as second right term in the eq. A.3 corresponding to the chemical potential of an electron in standard state $\mu°[e^-(G)] = F(4.44V - \chi_{aq})$ is obtained by substituting all these former numerical values in the eq. A.4.

The term χ_{aq} corresponds to the difference between the Galvani and Volta potentials, due to the internal charge distribution at the interfaces of different phases in which a surface potential is formed. The value of χ_{aq} is defined as 0.025V from [62].

Getting back to the eq A.3, the work function ϕ_{Metal} is defined as the difference between the chemical potential of an electron in standard state $\mu°[e^-(G)]$ and the real potential of an electron in the metallic phase $\alpha[e^-(Metal)]$.

$$F\phi_{Metal} = \mu°[e^-(G)] - \alpha[e^-(Metal)] \tag{A.5}$$

where the real potential of an electron in the metallic phase is defined as follows:

$$\alpha[e^-(Metal)] = \mu°[e^-(Metal)] - F\chi_{Metal} \tag{A.6}$$

where the factor χ_{Metal} represents the difference between Galvani and Volta potentials at the interphase of the metallic phase. Assuming the metallic phase as Platinum, the factors $\phi_{Pt} = 5.65V$ and $\chi_{Pt} = 15.9V$ according to [62]. Substituting these values in eq. A.3, the chemical potential of an electron in the Platinum can be expressed as follows:

$$\frac{\mu°[e^-(Pt)]}{F} = 14.66V \tag{A.7}$$

Now let us determine the cell potential for a PEMFC under reversible conditions. For the anode reaction:

$$H_2 \rightleftharpoons 2H^+ + 2e^- \tag{A.8}$$

the change in pure chemical free energy for the equilibrium state is:

$$\Delta G°_{anode} = -\mu°[H_2(G)] + 2\mu°[H^+(aq)] + 2\mu°[e^-(Pt)] \tag{A.9}$$

substituting the already known numerical values in the former equation $\Delta G°_{anode} = 2829,96$ kJmol^{-1} or $E°_{anode} = 14.66V$.

For the cathode reaction:

$$\frac{1}{2}O_2 + 2H^+ + 2e^- \rightleftharpoons H_2O \tag{A.10}$$

the change in pure chemical free energy for the equilibrium state is:

$$\Delta G°_{cathode} = \mu°[H_2O(L)] - \frac{1}{2}\mu°[O_2(G)] - 2\mu°[H^+(aq)] - 2\mu°[e^-(Pt)] \tag{A.11}$$

substituting the already known numerical values in the former equation and introducing new values like $\mu^\circ[O_2(G)] = 0$ and $\mu^\circ[H_2O(L)] = -237.18 \text{kJmol}^{-1}$ from [62], the change in pure chemical free energy for the cathode is $\Delta G^\circ_{\text{cathode}} = -3067.14 \text{kJmol}^{-1}$ which can also be expressed as follows $E^\circ_{\text{cathode}} = 15.89V$.

The PEMFC potential under reversible conditions at the standard state is:

$$E^\circ_{\text{PEMFC}} = E^\circ_{\text{cathode}} - E^\circ_{\text{anode}} = 1.23V \tag{A.12}$$

The ratio of oxidized to reduced molecules $[Ox]/[Red]$ is equivalent to the probability of being oxidized (giving electrons) over the probability of being reduced (taking electrons), which can be written in terms of the Boltzmann factor k for these processes as follows:

$$\frac{[Ox]}{[Red]} = \exp\left(\frac{\Delta G}{kT}\right) \tag{A.13}$$

which can also be expressed as

$$\Delta G = kT\ln\frac{[Ox]}{[Red]} \tag{A.14}$$

where k represents the Boltzmann constant.

If $[Ox]/[Red] = 1$, $\Delta G = \Delta G^\circ$. Therefore the former equation can also be written as

$$\Delta G = \Delta G^\circ + kT\ln\frac{[Ox]}{[Red]} \tag{A.15}$$

or

$$E = E^\circ + \frac{RT}{nF}\ln\frac{[Ox]}{[Red]} \tag{A.16}$$

For the PEMFC electrochemical mechanism the term

$$\frac{[Ox]}{[Red]} = \frac{[a_{H_2}][a_{O_2}]}{[a_{H_2O}]} \tag{A.17}$$

The cell potential can now be expressed as follows (general form of the Nernst equation under reversible conditions for different temperatures and concentrations):

$$E = E^\circ + \frac{RT}{nF}\ln\frac{[a_{H_2}][a_{O_2}]^{\frac{1}{2}}}{[a_{H_2O}]} \tag{A.18}$$

Assuming ideal gas conditions ($a_i = P_i/P_{\text{amb}}$), the former expression can be expressed as:

$$E = E^\circ + \frac{RT}{nF}\ln\frac{[P_{H_2}][P_{O_2}]^{\frac{1}{2}}}{[P_{H_2O}]} - \frac{RT}{nF}\ln P_{\text{amb}}^{\frac{1}{2}} \tag{A.19}$$

in practical applications of the Nernst equation (eq. A.18) the former expression eq. A.19 can also be presented as

$$E = 1.23 - 0.0009(T - 298) + \frac{RT}{2F}\ln(P_{H_2}P_{O_2}^{\frac{1}{2}}) \tag{A.20}$$

where the partial pressure of the water is suppressed for obvious reasons. Under irreversible (real) conditions, the Nernst equation can finally be expressed as follows:

$$E = 1.23 - 0.0009(T - 298) + \frac{RT}{2F}\ln(P_{H_2}P_{O_2}^{\frac{1}{2}}) - \eta_\circ \qquad (A.21)$$

where the factor $\eta_\circ$ represents the voltage drop due ohmic losses and reaction resistances.

Appendix B

Description of base models

B.1 Bernardi and Verbrugge, 1992

Steady state and isothermal 1D-Model applied in gas diffusion layers, polymer membrane and catalyst layers. Water management and the effect of membrane hydration on the PEMFC performance were the main objectives. They also studied the oxydant and fuel cross-over in Nafion and Dow membranes. They results focus on the transport of reactants and liquid water within the polymer itself as well as transport of reactants in the gas phase. They concluded that reactants are consumed completely within the first few microns of a 10 μm catalyst layer. Water transport across the membrane and catalyst layers is a function of electro-osmotic drag and applied pressure difference. The modeling results show that for an applied pressure difference of 2 atm, electro-osmotic drag can be offset for a current density of less than 0.6 A/cm2. In addition, the model is also used to illustrate the effect of GDL porosity on performance. Severe mass transport limitations are predicted for GDL porosities less than 20%. The model is well suited to simulating cell performance in regimes where mass transport limitations are not significant. In the regime where mass transport effects begin to dominate performance, the model does not agree well with experimental data. This is likely due to the fact that liquid water transport is not included in the model and to the assumption that reactants are transported only in dissolved form within the catalyst, as opposed to having a gas diffusion pathway.

B.2 Parthasaraty et al., 1992

Parthasarathy et al. [48] presented experimental results for many of the kinetic and mass transport parameters required for modeling the oxygen reduction reaction at the fuel cell cathode. They used a microelectrode, consisting of a thin platinum wire encased in Nafion, to simulate the catalyst layer of a proton exchange membrane fuel cell. They presented results for the Tafel slope, exchange current density, transfer coefficient, open circuit voltage, and solubility and diffusivity of oxygen in Nafion for temperatures ranging from 30 to 80°C. The tests were conducted with an applied oxygen pressure of 5 atm. The results for exchange current density show an increase in magnitude with temperature as well as with operating current. At low currents, there is an oxide layer present on the catalyst that reduces the exchange current density, this oxide layer is not present at high currents. The transfer coefficient also increases with temperature as does the open circuit voltage. The solubility of oxygen in Nafion decreases with temperature while the diffusion coefficient increases. The work does not include results showing how the solubility varies with local gas pressure and polymer water content.

B.3 Springer et al., 1991

One of the more important aspects of fuel cell operation is water management, which involves how water is transported within the cell. In their work, Springer et al. [2] explore how water is transported through the polymer membrane. Their paper has two sections. The first is the development of a computational model of a PEMFC that includes the gas diffusion layers and the membrane. The catalyst layers are treated as interfaces where source terms are applied. The second section includes experimental data for properties related to water uptake and transport through Nafion 117 membranes at 30°C. The water content of the membrane is given as a function of local vapor activity, and both the diffusion coefficient of water and the ionic conductivity of the membrane are given as a function of water content and temperature. In addition, a relation is derived that accounts for the amount of water transport by electro-osmotic drag per proton crossing the membrane from anode to cathode. Combining the experimental data with the computational model, Springer et al. generate cell performance curves for a range of operating conditions and also use the model to predict the net amount of water transport across the cell from anode to cathode. This turns out to be much less than what would be transported by electro-osmotic drag alone and indicates the importance of back diffusion from cathode to anode.

B.4 Zawodzinski et al., 1993

Zawodzinski et al. [15] presented an experimental study of water uptake and transport in Nafion 117 membranes and include comparisons with other types of ion exchange membranes. The results presented in this work are similar to those presented in Springer et al.. The authors show that Nafion 117 achieves a maximum water content of 14 H_2O/SO_3^- when in equilibrium with a saturated vapor at 30 °C and a maximum water content of 22 when immersed in liquid water at the same temperature (following pretreatment by boiling). The diffusion coefficient and ionic conductivity are also given as a function of membrane water content. In addition, the ionic conductivity of a fully immersed membrane is presented as a function of temperature as is data related to electro-osmotic drag.

Appendix C

Aspects of the numerical solution

C.1 Program structure

I Creation of text-file containing input variable names and magnitudes

    (a) Geometrical parameters

    (b) Electrochemical parameters

    (c) Operational parameters

II Subroutine-Start

    (a) Program reads the created input text-file and initializes every variable and array

    (b) Position vectors are created 3D (x,y and z vectors)

III Subroutine-Solution of straigth PEMFC numerical system

    (a) While-loop (Current density error>Tolerance)

    (b) Do i=1,..,m; Do j=1,..,n; Do k=1,..,p (focus on i,j,k node)

    (c) Depending on the domain i,j,k, the following subroutines run:

        i. Subroutine-Velocity field (Boundary conditions and iterative-discrete expression of momentum balance)

        ii. Subroutine-Mass transfer (Boundary conditions and iterative-discrete expression of specie mass balance)

        iii. Subroutine-Heat transfer (Boundary conditions and iterative-discrete expression of energy balance)

        iv. Subroutine-Two-phase flow in GDL (Boundary conditions and iterative-discrete expression of liquid water balance)

        v. Subroutine-Polymer membrane (Boundary conditions and iterative-discrete expression of water balance in polymer membrane)

    (d) Calculation of every distribution average-error

(e) Current density error > Tolerance?

    i. If No: Writing of 3D-arrays and other parameters in output text-files and END

    ii. If Yes: substitution of old values for calculated ones and new Iteration of While-loop

IV END

C.2 General scheme of the balance equations in discrete form

Calculation of ϕ-Property field. Combining integrated continuity equation and property balance in the finite volume-P assuming steady state conditions:

$$
\begin{aligned}
(J_{z+\Delta z} - F_{z+\Delta z}\phi_P) - (J_z - F_z\phi_P) \ + \\
(J_{y+\Delta y} - F_{y+\Delta y}\phi_P) - (J_y - F_y\phi_P) \ + \\
(J_{x+\Delta x} - F_{x+\Delta x}\phi_P) - (J_x - F_x\phi_P) \ = \\
(ST_C + ST_P\phi_P)\Delta x\Delta y\Delta z
\end{aligned}
\tag{C.1}
$$

where the factor J represents the convective and diffusive flow of the ϕ-Property as follows:

$$
J_i = \left(\rho V_i \phi_i - \Gamma \frac{\Delta \phi}{\Delta \text{Position}} \right) \Delta \text{Area}
\tag{C.2}
$$

the factor F represents the mass flow in eq. C.1

$$
\rho V_i \Delta \text{Area}
\tag{C.3}
$$

ST_C and ST_P represent the associated and non-associated source term, respectively. Doing the following nomenclature change:

$$
J_i - F_i \phi_P = a_i (\phi_P - \phi_{P+\Delta \text{Position}})
\tag{C.4}
$$

the final form of the eq. C.1 is expressed as follows

$$
a_P \phi_P = a_E \phi_E + a_W \phi_W + a_N \phi_N + a_S \phi_S + a_T \phi_T + a_B \phi_B + b
\tag{C.5}
$$

where

$$
a_P = a_E + a_W + a_N + a_S + a_T + a_B + a_P^\circ - ST_P \Delta x \Delta y \Delta z
\tag{C.6}
$$

and

$$
b = ST_C
\tag{C.7}
$$

the coefficient a is expressed as follows

$$
a_i = \frac{1}{2} \frac{\Gamma \Delta \text{Area}}{\Delta \text{Position}} \pm F_i
\tag{C.8}
$$

Appendix D

Data Validation

The input data used to validate the modeling and simulation in this work is presented as follows:

D.1 Data validation using Mench et al. 2003's experimental results

Table D.1: Geometrical and electrochemical parameters used as numerical input data (Mench et al. 2003)

Description	Value
channel height	0.00318 (m)
channel width	0.00216 (m)
channel length	0.14224 (m)
channel rib	0.00089 (m)
gas diffusion layer thickness	0.0002 (m)
gas diffusion layer porosity	0.6
catalyst layer thickness	10 (μm)
catalyst layer porosity	0.6
polymer membrane thickness	51 (μm)
Pt loading	0.5 $(mg_{Pt}cm^{-2})$
Pt surface area	400 (cm^2_{Pt}/mg_{Pt})
agglomerate porosity	0.26
agglomerate radii	0.00001 (cm)
Inlet Cathode Volumetric Flow	0.000001744 (m^3s^{-1})
Inlet Anode Volumetric Flow	0.00000079 (m^3s^{-1})
Inlet Cathode Flow Temperature	353.0 (K)
Inlet Anode Flow Temperature	363.0 (K)
Inlet Cathode Flow Pressure	150 (kPa)
Inlet Anode Flow Pressure	150 (kPa)

D.2 Data validation using Ju et al. 2005's experimental results

Table D.2: Geometrical and electrochemical parameters used as numerical input data (Ju et al. 2005)

Description	Value
channel height	0.00318 (m)
channel width	0.00216 (m)
channel length	0.14224 (m)
channel rib	0.00089 (m)
gas diffusion layer thickness	0.0002 (m)
gas diffusion layer porosity	0.6
catalyst layer thickness	10 (μm)
catalyst layer porosity	0.6
polymer membrane thickness	51 (μm)
Pt loading	0.5 ($mg_{Pt}cm^{-2}$)
Pt surface area	400 (cm^2_{Pt}/mg_{Pt})
agglomerate porosity	0.26
agglomerate radii	0.00001 (cm)
Inlet Cathode Volumetric Flow	0.000000821 (m^3s^{-1})
Inlet Anode Volumetric Flow	0.000001975 (m^3s^{-1})
Inlet Cathode Flow Temperature	353.0 (K)
Inlet Anode Flow Temperature	363.0 (K)
Inlet Cathode Flow Pressure	150 (kPa)
Inlet Anode Flow Pressure	150 (kPa)

References

[1] EGG Technical Services Inc Science application international corporation. *Fuel Cell Handbook*. U.S. Department of Energy, Office of Fossil Energy, National Energy Technology Laboratory, Morgantown, West Virginia 26507-0880, 6th edition, 2002.

[2] T. E. Springer, T. A. Zawodzinski, and S. Gottesfeld. Polymer electrolyte fuel cell model. *J. Electrochem. Soc.*, 138(8):2334–2342, 1991.

[3] S. Gottesfeld and T. A. Zawodzinski. *In Advances in Electrochemical Science and Engineering, Vol.5*. Wiley and Sons, New York, 1997.

[4] J. H. Nam and M. Kaviany. Effective diffusivity and water-saturation distribution in single- and two-layer pemfc diffusion medium. *Int. J. of Heat and Mass Transfer*, 46:4595–4611, 2003.

[5] M. Lampinen and M. Fomino. Analysis of free energy and entropy changes for half-cell reactions. *J. Electrochem. Soc.*, 140(12):3537–3546, 1993.

[6] D. Natarajan and T. Van Nguyen. A two-dimensional, two-phase, multicomponent, transient model for the cathode of a proton exchange membrane fuel cell using conventional gas distributors. *J. Electrochem. Soc.*, 148(12):A1324–A1335, 2001.

[7] T. Berning, D. M. Lu, and N. Djilali. Three-dimensional computational analysis of transport phenomena in a pem fuel cell a two-dimensional. *J. of Power Sources*, 106:284–294, 2002.

[8] Q. Wang, D. Song, T. Navessin, S. Holdcroft, and Z. Liu. A mathematical model and optimization of the cathode catalyst layer structure in pem fuel cells. *Electrochemical Acta*, 50:725–730, 2004.

[9] Chao-Yang Wang. Fundamental models for fuel cell engineering. *Chemical Reviews*, 104(10):4727–4766, 2004.

[10] J. Stumper, S. Campell, D. Wilkinson, M. Johnson, and M. Davis. In situ methods for the determination of current distributions in pem fuel cells. *Electrochimica Acta*, 43:3773, 1998.

[11] S. Cleghorn, C. Derouin, M. Wilson, and S. Gottesfeld. In situ current distribution measurements in polymer electrolyte fuel cells. *J. Appl. Electrochemistry*, 28:663, 2000.

[12] C. Wieser, A. Helmbold, and E. Guelzow. A new technique for two-dimensional current distribution measurements in electrochemical cells. *J. Appl. Electrochemistry*, 30:803–807, 2000.

[13] M. Noponen, T. Mennola, M. Mikkola, T. Hottinen, and P. Lund. Measurement of current distribution in a free-breathing pemfc. *J. of Power Sources*, 106:304–312, 2002.

[14] M. M. Mench and C. Y. Wang. An in situ method for determination of current distributionin pem fuel cells applied to a direct methanol fuel. *J. Electrochem. Soc.*, 150(1):A79–A85, 2003.

[15] T. A. Zawodzinski, C. Derouin, S. Radzinski, R. Sherman, V. T. Smith, T. Springer, and S. Gottesfeld. Water uptake by and transport through nafion 117 membranes. *J. Electrochem. Soc.*, 140(4):1041–1047, 1993.

[16] J. T. Hinatsu, M. Mizuhata, and H. Takenata. Water uptake of perfluorosulfonic acid membranes from liquid water and water vapour. *J. Electrochem. Sources*, 141(6):1493–1498, 1994.

[17] K. Broka and P. Ekdunge. Modeling the pem fuel cell cathode. *J. Applied Electrochemistry*, 27:281–289, 1997.

[18] K. Dannenberg, P. Ekdunge, and G. Lindbergh. Mathematical model of the pemfc. *Journal of Applied Electrochemistry*, 30(12):1377–1387, 2000.

[19] T. F. Fuller and J. Newman. Experimental determination of the transport number of water in nafion 117 membrane. *J. Electrochem. Soc.*, 139(5):1332–1337, 1992.

[20] W. Neubrand. *Modellierung und Simulation von Electromembranverfahren.* Dissertation-Universität Stuttgart, Germany, 1999.

[21] D. M. Bernardi and M. W. Verbrugge. Mathematical model of the solid polymer electrolyte fuel cell. *J. Electrochem. Soc.*, 139(9):2477–2497, 1992.

[22] V. Gurau, H. Liu, and S. Kakac. Two-dimensional model for proton exchange membrane fuel cells. *AIChE J, 44, 2410-2422*, 44:2410–2422, 1998.

[23] J. Yi and T. V. Nguyen. A two-phase flow model to investigate the hydrodynamics of gas and liquid water in the cathode of pem fuel cells with interdigitated gas distributors. *AIChE Meeting Miami Beach*, Paper107b:15–20, 1998.

[24] J. Yi and T. V. Nguyen. Multi-component transport in porous electrodes in proton exchange membrane fuel cells using the intedigitated gas distributors. *J. Electrochem. Soc.*, 146(38), 1999.

[25] S. Um, C. Y. Wang, and K. S. Chen. Computational fluid dynamics modeling of proton exchange membrane fuel cells. *J. Electrochem. Soc.*, 147(12):4485–4493, 2000.

[26] C. Y. Wang and P. Cheng. A multiphase mixture model for multiphase, multicomponent transport in capillary porous media-i: model development. *Int. J. Heat Mass Transfer*, 39:3607–3618, 1996.

[27] S. Dutta, S. Shimpalee, and J. W. Van Zee. Three-dimensional numerical simulation of straight channel pem fuel cells. *J. Applied Electrochemistry*, 30:135–146, 2000.

[28] S. Dutta, S. Shimpalee, and J. W. Van Zee. Numerical prediction of mass-exchange between cathode and anode channels in a pem fuel cell. *Int. J. Heat Mass Transfer*, 44:2029–2042, 2001.

[29] T. Berning and N. Djilali. A 3d multiphase, multicomponent model of the cathode and anode of a pem fuel cell. *J. Electrochem. Soc.*, 150(12):A1589–A1598, 2003.

[30] Z. H. Wang, C. Y. Wang, and K. S. Chen. Two-phase flow and transport in the air cathode of proton exchange membrane fuel cells. *J. of Power Sources*, 94:40–50, 2001.

[31] D. Natarajan and T. Van Nguyen. Three-dimensional effects of liquid water flooding in the cathode of a pem fuel cell. *J. Power Sources*, 115:66–80, 2003.

[32] L. You and H. Liu. A two-phase flow and transport model for the cathode of pem fuel cells. *Int. J. Heat Mass Transfer*, 45:2277–2287, 2002.

[33] U. Pasaogullari and C-Y Wang. Liquid water transport in gas diffusion layer of polymer electrolyte fuel cells. *J. Electrochem. Soc.*, 151:A399–A406, 2004.

[34] U. Pasaogullari and C-Y Wang. Two-phase modeling and flooding prediction of polymer electrolyte fuel cells. *J. Electrochem. Soc.*, 152(2):A380–A390, 2005.

[35] H. Meng and C. Y. Wang. Model of two-phase flow and flooding dynamics in polymer electrolyte fuel cells. *J. Electrochem. Soc.*, 152(9):A1733–A1741, 2005.

[36] F. Y. Zhang, X. G. Wang, and C. Y. Wang. Liquid water removal from a polymer electrolyte fuel cell. *J. Electrochem. Soc.*, 153:A225–A232, 2006.

[37] G. Guvelioglu and H. Stenger. Computational fluid dynamics modeling of polymer electrolyte membrane fuel cells. *J. of Power Sources*, 147:95–106, 2005.

[38] K. Lum and J. McGuirk. Three-dimensional model of a complete polymer electrolyte membrane fuel cell-model formulation, validation and parametric studies. *J. of Power Sources*, 143:103–124, 2004.

[39] S. Um and C. Y. Wang. Three-dimensional analysis of transport and electrochemical reactions in polymer electrolyte fuel cells. *J. of Power Sources*, 125:40–51, 2004.

[40] H. Meng and C. Y. Wang. Large-scale simulation of polymer electrolyte fuel cells by parallel computing. *Chemical Engineering Science*, 59:3331–3343, 2004.

[41] J. Ihonen, F. Jaouen, G. Lindbergh, and G. Sundholm. A novel polymer electrolyte fuel cell for laboratory investigations and in-situ contact resistance measurements. *Electrochemical Acta*, 46:2899–2911, 2001.

[42] M. M. Mench, C. Y. Wang, and M. Ishikawa. In situ current distribution measurements in polymerelectrolyte fuel cells. *J. Electrochem. Soc.*, 150(8):A1052–A1059, 2003.

[43] H. Ju, C-Y Wang, S Cleghorn, and U Beuscher. Nonisothermal modeling of polymer electrolyte fuel cells:i experimental validation. *J. Electrochem. Soc.*, 152(8):A1645–A1653, 2005.

[44] Y. Wang and C-Y Wang. Modeling polymer electrolyte fuel cells with large density and velocity changes. *J. Electrochem. Sources*, 152(2):A445–A453, 2005.

[45] M. L. Perry, J. Newman, and E. J. Cairns. Mass transport in gas-diffusion electrodes: A diagnostic tool for fuel-cell cathodes. *J. Electrochem. Soc.*, 145(1):5–15, 1998.

[46] C. Marr and X. Li. Composition and performance modeling of a catalyst layer in a proton exchange membrane fuel cell. *J. of Power Sources*, 77:17–27, 1999.

[47] F. Jaouen, G. Lindbergh, and G. Sundholm. Investigation of mass-transport limitations in the solid polymer fuel cell cathode. *J. Electrochem. Soc.*, 149(4):A437–A447, 2002.

[48] Parthasarathy, S. Srinivasan, and J. Appleby. Temperature dependence of the electrode kinetics of oxygen reduction at the platinum/nafion interface - a microelectrode investigation. *J. Electrochem. Soc.*, 139(9):A437–A447, 1992.

[49] M. M. Tomadakis and S. V. Sotirchos. Ordinary and transition regime diffusion in random fiber structures. *AIChE J.*, 39:397–412, 1993.

[50] R. B. Bird, W. E. Stewart, and E. N. Lightfoot. *Transport Phenomena*. Wiley, New York, 1960.

[51] D. Chung and R. Jiang. Performance of polymer electrolyte membrane fuel cell (pemfc) stacks. *J. Power Sources*, 83:128–133, 1999.

[52] S. Morner and S.A. Klein. Experimental evaluation of the dynamic behavior of an air-breathing fuel cell. *Journal of Solar Energy Engineering*, 123:225–231, 2001.

[53] M. Noponen, T. Mennola, M. Mikkola, T. Hottinen, and P. Lund. Determination of mass diffusion overpotential distribution with flow pulse methodfrom current

distribution measurements in a pemfc. *J. Appl. Electrochemistry*, 32:1081–1089, 2002.

[54] T. Mennola, M. Mikkola, M. Noponen, T. Hottinen, and P. Lund. Measurement of ohmic voltage losses in individual cells of a pemfc stack. *Journal of Power Sources*, 112:261–272, 2002.

[55] P.W. Li, T. Zhang, Q. M. Wang, L. Schaefer, and M. K. Chyu. The performance of pem fuel cells fed with oxygen through the free-convection mode. *Journal of Power Sources*, 114:63–69, 2003.

[56] T. Hottinen, M. Noponen, T. Mennola, O. Himanen, M. Mikkola, and P. Lund. Effect of ambient conditions on performance and current distribution of a polymer electrolyte fuel cell. *Journal of Applied Electrochemistry*, 33:265–271, 2003.

[57] T. Mennola, M. Noponen, M. Aronniemi, T. Hottinen, M. Mikkola, O. Himanen, and P. Lund. Mass transport in the cathode of a free-breathing polymer electrolyte membrane fuel cell. *Journal of Applied Electrochemistry*, 33:979–987, 2003.

[58] A. Schmitz, S. Wagner, R. Hahn, H. Uzun, and C. Hebling. Planar self-breathing fuel cells. *Journal of Power Sources*, 118:162–171, 2003.

[59] T. Mennola, M. Noponen, T. Kallio, M. Mikkola, and T. Hottinen. Water balance in a free-breathing polymer electrolyte membrane fuel cell. *Journal of Applied Electrochemistry*, 34:31–36, 2004.

[60] A. Schmitz, S. Wagner, R. Hahn, H. Uzun, and C. Hebling. Stability of planar pemfc in printed circuit board technology. *Journal of Power Sources*, 127:197–205, 2004.

[61] T. Hottinen, M. Mikkola, and P. Lund. Evaluation of planar free-breathing polymer electrolyte membrane fuel cell design. *Journal of Power Sources*, 129:68–72, 2004.

[62] JANAF. *Thermodynamical Tables*. Supplement 1975. 2nd edition, 1971.

[63] J.R. Farrel and P. McTigue. Precise compensating potential difference measurements with a voltaic cell. *J. Electroanal. Chem.*, 139(37), 1982.

Prepublications of parts of this work

(1) L. Matamoros and D. Brüggemann, Simulation of the water and heat management in proton exchange membrane fuel cells, Journal of Power Sources, 161, 203-213, 2006.

(2) L. Matamoros and D. Brüggemann, Numerical study on PEMFC's geometrical parameters under different humidifying conditions, Journal of Power Sources, 172, 253-264, 2007.

(3) L. Matamoros and D. Brüggemann, Concentration and ohmic losses in free-breathing PEMFC, Journal of Power Sources, 173, 367-374, 2007.

In der Reihe „*Thermodynamik: Energie, Umwelt, Technik*", herausgegeben von Prof. Dr.-Ing. D. Brüggemann, bisher erschienen:

ISSN 1611-8421

1	Dietmar Zeh	Entwicklung und Einsatz einer kombinierten Raman/Mie-Streulichtmesstechnik zur ein- und zweidimensionalen Untersuchung der dieselmotorischen Gemischbildung	
		ISBN 978-3-8325-0211-9	40.50 €
2	Lothar Herrmann	Untersuchung von Tropfengrößen bei Injektoren für Ottomotoren mit Direkteinspritzung	
		ISBN 978-3-8325-0345-1	40.50 €
3	Klaus-Peter Gansert	Laserinduzierte Tracerfluoreszenz-Untersuchungen zur Gemischaufbereitung am Beispiel des Ottomotors mit Saugrohreinspritzung	
		ISBN 978-3-8325-0362-8	40.50 €
4	Wolfram Kaiser	Entwicklung und Charakterisierung metallischer Bipolarplatten für PEM-Brennstoffzellen	
		ISBN 978-3-8325-0371-0	40.50 €
5	Joachim Boltz	Orts- und zyklusaufgelöste Bestimmung der Rußkonzentration am seriennahen DI-Dieselmotor mit Hilfe der Laserinduzierten Inkandeszenz	
		ISBN 978-3-8325-0485-4	40.50 €
6	Hartmut Sauter	Analysen und Lösungsansätze für die Entwicklung von innovativen Kurbelgehäuseentlüftungen	
		ISBN 978-3-8325-0529-5	40.50 €
7	Cosmas Heller	Modellbildung, Simulation und Messung thermofluiddynamischer Vorgänge zur Optimierung des Flowfields von PEM-Brennstoffzellen	
		ISBN 978-3-8325-0675-9	40.50 €
8	Bernd Mewes	Entwicklung der Phasenspezifischen Raman-Spektroskopie zur Untersuchung der Gemischbildung in Methanol- und Ethanolsprays	
		ISBN 978-3-8325-0841-8	40.50 €